What people are sayin

M000024986

"Now that's innovative - a business book that's entertaining too. The powerful lessons that Dalton shares inside are like jet fuel for companies that want to propel their growth."

—Ed Petkus, VP - New Products & Engineering,
Hawker-Beechcraft Aviation

"Dalton's innovative approach can help any company find the 80/20 spike needed to drive new product growth."

—Richard Koch, bestselling author of *The 80/20 Principle*

"*Simplifying Innovation* offers an elegant solution for any company trying to get more out of their R&D investment. A refreshingly original approach that is a must for any executive's library."

—Matthew E. May, author of *In Pursuit of Elegance*
and *The Elegant Solution*

"A Theory of Constraints approach to the process of innovation was long overdue. Production, project management, supply chain, and policy constraint analysis have all been comprehensively addressed. But until now, no one has thought to examine the ramifications of constraint theory on the challenge of innovation. Mike Dalton's novel was worth waiting for. *Simplifying Innovation* synthesizes innovation best practices and the focusing step framework to create a powerful new application of TOC. Let it stimulate your imagination as it did mine."

—H. William Dettmer, author of *Strategic Navigation*

"Organic growth is hard work, but there is much you can learn from Mike Dalton's years of practical experience. If you are committed to maintaining the discipline required, *Simplifying Innovation* offers a refreshing, common sense approach for improving your new product results and creating profitable growth. I'd recommend it for any business leader's bookshelf."

—Dr. James D. Hlavacek, author of
Profitable Top Line Growth for Industrial Companies

"Inside a fascinating business novel, that I literally couldn't put down, Mike Dalton has created a hands-on field manual to extending the Theory of Constraints to innovation - I only wished I had the benefit of Mike's insights during my days as an R&D leader in Bell Labs."

—Dr. Matthew W. Sagal, co-author of *The Strongest Link*

Simplifying Innovation

Doubling speed to market and new product profits—with your existing resources

Michael A. Dalton

Flywheel Effect
Publishing

Division of Guided Innovation Group, LLC

Simplifying Innovation

Doubling speed to market and new product profits—with your existing resources

Michael A. Dalton

Special bulk quantity discounts are available to corporations, professional service associations, and other organizations. For details, please contact:
publisher@flywheeleffect.com

Disclaimer:

Library of Congress Cataloging-in-Publication Data applied for
ISBN- 978-0-615-32939-0

"Three Rules of Work: Out of clutter find simplicity; From discord find harmony; In the middle of difficulty lies opportunity."

—*Albert Einstein*

CONTENTS

360 Degrees of Complexity

Whether you are a C-level executive, a group manager responsible for R&D or marketing, or a project manager responsible for a new product development team, you face unprecedented complexity and pressure in your business today:

- From customers demanding new solutions, higher quality, lower costs, and wider availability
- From global competitors intent on disrupting markets
- From employees expecting unprecedented flexibility, benefits, advancement, and compensation
- From governments playing a wider role in how you operate

And let's not forget shareholders, whose expectations for growth are increasingly difficult to achieve, given all of the other demands you face. With that kind of pressure, it feels like there is no end in sight.

Of course, it shouldn't be a surprise that the expectations are so high. On average, companies that depend on new products for growth spend 3.6% of net sales on R&D– sometimes several times that in new technology markets. However, most companies struggle to see the new product growth they need from that investment. In fact, 50% of CEO's rate their innovation return as unsatisfactory.[1]

For most companies, that means a stunning 35% or more of net income goes toward R&D, without delivering the level of growth needed to meet those rising expectations. Even more startling is the fact that we squander over $75 billion dollars on innovation that doesn't deliver every year in the U.S. manufacturing sector alone.[2]

"Why should anyone be struggling to achieve improved returns?" you may wonder. After all, there are so many strategies available: There's open innovation, disruptive innovation, customer-focused innovation, business model innovation, strategic innovation, game-changing innovation, outcome-driven innovation, agile innovation, voice of customer, quality function deployment, lean product development...Well, you get the idea.

If you're like most companies, you've tried some of these approaches, maybe even with good results initially, only to see them go by the wayside when the next big idea comes along. There are so many concepts out there and so many levers to pull that it's easy to become enamored with a new approach, only to find it doesn't fit your situation as well as you thought it might. So, how do you make sense of this complexity, choose the ideas that will have your desired effect, and then move from concept to execution in order to develop the innovation culture you are seeking?

The fact of the matter is that innovation will always be complex. However, making improvements to it doesn't have to be. Year after year, companies squeeze extra productivity out of their manufacturing and even their service organizations. Yet, few know how to do the same thing with their innovation process—systematically extracting more and more from the innovation resources they already have. While we seldom hear about them, such companies do exist. Even with the issues it faces today, Toyota is one example of a company that has continually improved its innovation capabilities. Even with R&D spending significantly below most of its peers, it manages to outperform the competition.[3]

In fact, the single most powerful way to drive new product improvement is to implement a systematic framework for improvement. Using a time-tested approach will help you focus your efforts and drive continuous improvement in your new product process: improvement that will increase profits and reduce the delays and cost overruns that plague companies everywhere.

Simplifying Innovation reveals how you can take the complex subject of innovation and use a simple, high-leverage approach to drive continuous improvement of your new product results. It's based on Theory of Constraints (TOC)—a proven and well-established approach for manufacturing improvement that is now being used to systematically deliver innovation gains with the new product development resources that companies already have in place.

Companies have realized gains ranging:

- Up to a 50% reduction in new product development time, from start to finish
- As high as a 100% increase in net profits on new products
- As much as a 20% reduction in overall development expenditures

Throughout the following chapters, you'll be introduced to Barrister Industries and the new general manager of one of its divisions, Maggie. A fictional company struggling to get its new product innovation working again, Barrister's story demonstrates the issues companies like yours face every day and how a systematic framework can help you address them. While the underlying problems that Barrister faces may be complex, through them, you'll learn how TOC can simplify any situation. You'll also learn a straightforward, five-step approach for identifying the bottleneck in your new product process, uncovering the core issues that constrain it, and engaging your team to drive improvements in new product development speed and impact.

Don't worry about trying to keep track of the details as the story progresses. Throughout key chapters, you'll receive enough information to reveal how the process unfolds.

Inspired by Eli Goldratt's approach in *The Goal*, the business novel format takes readers beyond the words and into an experience of the framework. However, some readers prefer a more traditional approach, so we included an overview at the end. Part V summarizes all of the elements of the story into a concise outline of the entire system and the keys to using it within your organization—what we call the Guided Innovation System™: a high-leverage, step-by-step process that you can implement immediately to begin getting more profits from your new product investment.

Some readers may prefer to read the summary in Part V first and then refer to it throughout the story. If you'd like a printout for making notes, you can download a copy of the summary as well as other useful resources at:

www.SimplifyingInnovation.com/extras

Special Note for the TOC Community

For my colleagues in the field of Theory of Constraints, you will undoubtedly find that the first half of this book covers material that you are already quite familiar with but is important background for the majority of readers who will be new to TOC.

To help establish realistic expectations, I didn't write *Simplifying Innovation* to reflect the leading edge of TOC thinking. Rather, I wrote it to reflect the synthesis of TOC fundamentals with the latest in new product development and marketing concepts and practices. My purpose is to advance the art and science of innovation management so that companies can get more impact from their R&D investment—to bring the power of the TOC thinking processes to helping companies focus and select the right tools for driving improvement in speed to market and new product profits.

I hope you'll join me in recognizing the value of expanding the application of TOC and in welcoming others into the TOC community. I hope that *Simplifying Innovation* inspires you to pursue further applications as well.

For those of you who are new to TOC, but are intrigued and want to learn more, I would direct you to the Theory of Constraints International Certification Organization. The TOCICO is a global not-for-profit certification organization for TOC practitioners, consultants and academics. It develops and administers certification standards and facilitates the exchange of the latest developments in the TOC body of knowledge. Their website is www.TOCICO.org

The Story

Part I - Seismic Shift

Epicenter

Maggie Edwards sat in stunned silence with the rest of the shell-shocked executive team. It was 4:30 on a Friday afternoon and they were scattered around a conference table in the large, modern stainless and glass office of Doug Stanton, President and CEO of Barrister Industries—the office where J. Randolph Barrister III, the company's third generation chairman and largest shareholder, had just dropped the bomb.

"Thank you all for coming, and let me get right to the point. Barrister Industries has just signed a letter of intent to sell the TerraGrafix division, your division, to a leading competitor. We will be announcing the agreement Monday and, barring unforeseen problems, we intend to close the sale to Globalgraf 30 days from now."

Looking around the conference table at the rest of the executives, Maggie saw that everyone else had the same look of disbelief. Her mind was racing, so much so that she barely heard anything else he said. *What's going to happen to everyone I know? Why would they let go of one of the most profitable companies in the Barrister portfolio? How did they manage to keep this quiet?* On second reflection, the division controller didn't seem surprised at all and was instead looking down at his lap. He had probably been carrying this burden for some time, providing the necessary financial information required for due diligence.

The CEO took over from there. "Thanks, Randy. As you already know, this has been a difficult year for Barrister Industries. Several of our core divisions were heavily dependent on automotive, and I'm afraid that the recent downturn has been an enormous drain on cash flow." Shaking his head with a genuine look of humility, Doug said, "Far worse than I could have ever expected."

"While TerraGrafix has been a gem in the Barrister portfolio, I'm sorry to say that we are being forced to take this step to keep the entire company solvent. For the greater good, you might say—as little solace as that might be to all of you."

It was all becoming clear now. Randy Barrister had been determined to take the company to a higher level than previous generations had ever dreamed. Initially, his acquisition strategy had gone well, but several underperforming acquisitions and mounting debt obligations had begun to unravel his vision. Now, with the economic slowdown, those obligations had become stifling. TerraGrafix was one of the only divisions performing well enough to provide the cash infusion needed. Barrister was selling off one of its prize cows to save the rest of the farm.

"The leadership at Globalgraf is committed to retaining as much of the staff as possible. Therefore, we will be setting up interviews in the coming weeks so that their integration team can get to know your people and their capabilities. While I can't say for sure what that means for all of you as part of the executive team, I can reassure you that we have set aside a generous reserve for transitional retention and severance."

There it was. The folks in this room weren't born yesterday. They knew that many of their colleagues around the table and some of the people who worked for them would become 'synergies,' a euphemism the merger and acquisition trade used for firing people in redundant job positions.

After all, Globalgraf wouldn't need another Manufacturing VP. No, there would be some sort of interim agreement where Maggie would help to transition products to the new facility, but eventually, she would be out of a job—with a severance package, but still out of a job. That was something Maggie hadn't faced since she had graduated from engineering school, almost 20 years ago.

During the next 20 minutes, Doug went on to explain what would be happening in the next few weeks and how Globalgraf wanted to handle employee and customer communications. After taking questions, he adjourned the meeting. While the rest of the battered executive team exited the room, he pulled Maggie aside. "Can I ask you to stay for a few minutes longer?"

"Umm...Sure, Doug," she replied. *What could this be about?*

Attenuation

As the last of the group filed out, Maggie stood in the small reception area outside of Stanton's office, staring at a display of products from each of the divisions. Several minutes later, Randy Barrister emerged from Doug's office, nodded uncomfortably in Maggie's direction, and without making any eye contact, hurried down the hallway.

Doug stepped out to welcome her. "Come on in, Maggie." Then, closing the door, he gestured toward two chairs at the corner of the large conference table. "Please, take a seat," he said. After they were seated, he continued, "Maggie, first off, let me say that I feel just terrible about what we've had to do. I want you to know that we really appreciate all you've done for us over the years at TerraGrafix."

Well, you certainly have an odd way of showing it, she felt like saying, but just smiled.

"In a growth industry, you've done an amazing job of delivering increased capacity year in and year out, as well as managing the addition of new capacity."

Normally, she might have been flattered, but today was anything but normal. *Where is he heading with this?*

"As I hope you can appreciate," he said, "I can't get in the middle of things right now between you and Globalgraf, but I know you're sharp enough to have already figured out that they may not need another head of manufacturing."

She nodded. *So my speculation wasn't too far off.*

"Well, their leadership is pretty sharp, too, so I expect that they'll try to find some kind of position for you. But if they don't have an opportunity that interests you, we should talk."

"What do you have in mind?"

"Direct and to the point, Maggie, that's part of what I've always respected about you," he smiled. "While I'm afraid I don't have a manufacturing leadership role that I could offer you, I'm going to need someone that could lead a product development push. I can't really say too much more. Just promise me that you'll come and see me if they don't have a role for you after your transition obligations wind down."

Unsure of what to think, but knowing that it was best to leave all her options open, Maggie agreed to stay in touch. She made her way out of the headquarters building and got into her car for the short commute home.

What was Doug smoking? she wondered as she accelerated out of the parking garage. *I'm a manufacturing executive, not a PhD. What in the world do I know about creating new products?* Besides, he'd just sold the division that she had spent years helping to build. Why should she want to help him? She shook her head, *No, it's not right to think that way.* Doug was simply doing what he had to do in order to protect the company and save thousands of jobs in the other divisions.

Realizing she'd been driving far too aggressively, she took a deep breath to calm herself. As she entered the historic district where she and her family had recently purchased and begun restoring a rambling, old turn-of-the-century three story, her thoughts turned to how she was going to break the news to them...

Salvage operations

The sale of TerraGrafix stormed ahead faster than anyone expected. In her usual style, Maggie was so busy and involved in ensuring a successful transition that she had nearly forgotten the CEO's invitation to come back and talk with him. But as Doug Stanton had predicted, the Globalgraf team was sharp. They recognized Maggie's talent and energy, and two weeks before the end of the transition period, George Malone met with her to offer her a new position.

"Maggie, Globalgraf can use someone like you, but I understand that you're not interested in relocating to our headquarters." Maggie swallowed, then nodded in agreement.

"Unfortunately, that means we can't offer you a position heading up manufacturing. But, we still think that you can play a key role in our organization." Maggie leaned forward, showing her interest.

"Globalgraf needs someone who can ensure that each plant learns from and applies industry standard best practices. We think you'd be a perfect fit." He smiled warmly as Maggie waited silently.

"The position would be at your current level and would allow you to stay on without relocating. Of course, there would be a little bit of travel. You'd have to visit the various plant locations approximately three weeks out of every month."

Three weeks is a little travel. I'd hate to hear what you think is a lot of travel, she almost groaned.

"So tell me, Maggie. What do you think?"

Searching for a way to buy more time she said, "It's a very kind offer, George, and sounds like quite an interesting role. Of course, it would be a big change, so I'm sure you'll understand that I have to talk it over with my family."

Looking a little surprised, George said, "Oh yes, please take some time to discuss it at home."

After they had said their goodbyes, Maggie sat down and reflected quietly in her office. *This isn't going to go over very well at home...*

False start

By the time Maggie helped her twin nine-year-olds, Luke and Leo, finish their math homework, it was time to send them to bed. With the twins settled, she stopped and said goodnight to her twelve-year-old daughter, who was finishing homework in her room while simultaneously engaging in a rapid-fire texting conversation. Sophia had always been a cute kid, with the same auburn hair and fair complexion as her mother and grandfather. But, she also had her father's deep blue eyes, and Maggie realized what a striking young woman she was becoming.

"Don't stay up too late, Sophia," she said.

Of course, that elicited the slightest of eye rolls, as if to say, "Mom, I'm not a little girl anymore." But, "Okay, Mom," was the dutiful reply. She was getting to that difficult age, but she was still a good kid.

Finally, with everyone tucked in safely for the night, Maggie plopped into bed next to her husband, Jeff. Holding a glass of wine, she let out the faintest of sighs.

"Okay, Maggs, out with it. You've been distracted all night. What's up?"

She smiled at how well he could read her. "Do you remember that conversation we had about my meeting with Doug Stanton at Barrister?"

He chuckled. "How could I forget? Hi. Your division has been sold. Don't let the door hit you on the way out." It was only natural that Jeff wanted to defend his wife. As her husband, he wasn't exactly going to be sympathetic to Barrister Industries' situation.

"Jeff, it wasn't quite like that. Anyway, I'm referring to the part where he said that he was sure Globalgraf would try to create a role for me."

"But Maggie, we already agreed that we're not relocating. I have tenure in the school district, and with Sophia in middle school..."

"No disagreement," she interrupted, putting her hands up in mock surrender. "Actually, they've offered me a challenging new position that allows me to stay here in town."

"But...?" he coaxed, as if he was reading her mind.

"But it would include some travel. Well, actually lots of travel—as much as three weeks a month."

"Is that really what you want?"

"Well, I really like the idea of helping all the different locations improve, and we do have a lot of expenses. But, I just can't see how a life jetting back and forth between plant locations would be good for us."

Relief seemed to spread across Jeff's face for a brief moment before he asked, "What about your other options? Couldn't you talk to some of other companies here in town?"

"Jeff, with the economy the way it is, most manufacturers are pulling back. Plus, when I step back and look at it, anything I did find would probably just be more of the same thing that I've been doing for years now."

"Maybe you could take the Globalgraf position, just to stay busy while you look for something else."

"Yeah, but you know me. I'm worried I'd get too busy and wouldn't spend enough time looking outside. Before you know it, we'd be a year into living apart."

"Yeah, I've seen that enough times before," he smiled. "Maybe you could just take some time off and supervise the contractors here at the house."

"That would be great, but you know what this is costing." They had maxed out their credit line with a mortgage on the house and then a second for the authentic restoration they had always dreamed of doing. "Besides, at the least, I'd like to be able to send our kids to junior college," she laughed, remembering the commitment they'd made long ago to make sure the kids got a top quality education.

"I'm sure the contractors will be relieved to hear that," he teased.

"Hey! Take that back," she said, punctuating the request with an elbow to his ribs.

Okay, okay... I know you'll think of something, Maggs."

And with that, she snuggled into his chest to think about her options.

Fresh perspective

The next morning, Maggie woke early, resigning herself to the fact that there was far too much weighing on her mind to sleep. Jeff didn't stir as she put on her workout gear, but their rescue American Pitbull Terrier, Sweetie, opened up one eye and yawned as if to say, "Really, it's not even light yet."

After making her way down to the kitchen, her eyes took in the room and she realized for at least the hundredth time how much she liked the way it had turned out. While their goal was to do an authentic restoration of the old house, the kitchen was one of the first rooms they'd tackled, deciding they would detour from their plan by adding all of the modern conveniences, including a large center island. The result was a warm, cheery décor, which had turned into an inviting family hub.

Unfortunately, it was too early for the automatic coffee maker, so Maggie pulled on her running shoes and stepped outside with Sweetie in tow. When the now fully awake dog was finished with her business and had done a quick squirrel patrol, they took off on their run. She really enjoyed running in the cool morning air, with the muscular dog loping along by her side.

Forty minutes later as they made their way back up the driveway, Maggie realized that she didn't remember any part of the run. That's what she had hoped for—that focused feeling she got from running that automatically allowed her mind to go off and process things on its own.

As she stepped inside the house, she inhaled the aroma of freshly brewed coffee. The exhausted dog, recognizing that the smell wasn't food, quickly laid down for a nap on the kitchen rug. Maggie tucked the morning newspaper under her arm and poured a cup of coffee in her favorite mug. It was the one Sophia had made in a craft class at summer camp back before her parents had become a point of embarrassment. If what her friends had all told her about adolescence was true, the next few years would feel like they were as long as the first twelve.

Maggie hopped onto the stool at the island counter. As she sipped the intensely dark brew, she was reminded of childhood: of coming downstairs early to find her father having dry toast and coffee before heading to his shift at the mill. As a steelworker, he'd managed to give them a good life, but it hadn't been easy. As American industry struggled, he vowed his children would get the university education he didn't have. Joseph Sullivan had beamed when his daughter Maggie graduated near the top of her class in engineering school and then again when she'd gotten her masters in business. He was especially proud because she had insisted on working at an assembly plant to help pay for the first two years of undergraduate school and had become a paid co-op student after that. He had died two years ago, and as she sat in the kitchen she still needed to pay for, she wondered what he would have thought about her situation.

As she sipped her coffee, Maggie finally realized that she had been ignoring the offer that Doug had made. "I guess it wouldn't do any harm to talk," she said aloud. Sweetie opened an eye to assess whether the remark was directed at her and, of course, whether it might be accompanied by food. As Maggie opened up the paper, she resolved to share Stanton's offer of sorts with Jeff when he came downstairs. She knew he'd still be skeptical about Barrister Industries. But she also knew he'd agree that it couldn't hurt to talk. It was time to hear what Doug had to offer.

Aftershocks

"Thanks for fitting me into your busy schedule, Doug," said Maggie as she took a seat again near the corner of the large conference table.

"Maggie, I'd almost given up hope that you would call. Can I offer you some coffee?" he asked, gesturing to the tray that his secretary brought in before Maggie arrived.

"Please. That would be great."

After pouring two cups, he smiled and continued. "I thought that Globalgraf might have come to their senses and made you an offer that was too good to refuse."

She wasn't quite sure what to think about Doug, but she launched in anyway. "They did, but...honestly, I've already told them that I'm not going to take it. There was just too much travel involved."

"I'm sure they're disappointed," he said with sincerity mixed with a tinge of satisfaction. "Up until now, I couldn't say much because I had an obligation to let Globalgraf make you an offer first. But, it would seem that's behind us now. Maggie, the reason I wanted to talk is that I have a real product development mess in the Dynamic Fluid Technologies division."

With the sale of TerraGrafix, Dynamic Fluid Technologies was the second largest division of Barrister Industries. DFT's core business was products for water treatment and filtration, ranging from small home filtration units all the way up to large municipal and industrial treatment and measurement systems.

"I need someone to help get new products moving again."
"Okay..." Maggie wasn't sure where Doug was heading. She certainly didn't have any experience in product development.

"Maggie, I think you'd be a perfect fit, and I'd like to offer you the position of Executive Vice-President and General Manager for the DFT division with a nice bump in salary and the next level in bonus, as well. Roger Huntley, the current VP and GM, is being moved to the position of business development where he will report to me."

Maggie hadn't really known what to expect today and had thought they might just end up kicking around some possibilities. She certainly hadn't expected this.

"How's Roger going to respond?"

"He's had a year to act, Maggie, and frankly, his only response has been to crack the whip and place blame. We both know that doesn't work."

"But, Doug..."

Raising his hand, he stopped her. "You're going to tell me that you don't have any experience with new products. I know that already, but I'm not concerned. I'm looking for a fresh perspective. I know you offer that. Not only that, but you know the ropes at Barrister. Most importantly, you have a track record that demonstrates your innate ability to improve processes."

"But product development is more of an art, isn't it?"

"At least that's what the folks around here are always telling me," said Doug. "Frankly, though, I'm sick and tired of hearing that we can't ask for more—that increasing our organic growth and getting more out of our new products is going to require increasing our investment. At the other end of the spectrum, there are those on the leadership team who think it's hopeless and would like to slash R&D spending or outsource it completely."

Her radar went up. "You're not asking me to swing the hatchet, I hope! That's not what I'm about."

"Easy there, Maggie...Again, I respect your directness, I really do, but that's not at all what I was thinking," Doug said, quickly putting her at ease. "No, there are two contradictory assumptions here. One is that we need to cut R&D costs to be more profitable. The other is that we need to get more out of our R&D investment. I want to see what we can do to prove the second one."

Maggie's curiosity picked up, "Well what kind of problems are you seeing?"

"Honestly, I'm not thrilled with the new product growth in any of our divisions, but the delays at DFT have become unacceptable," he grimaced. "Not a single project is on time, every program is over budget, and less than half of their projects make it to market. And those that do, well they rarely meet their sales projections."

"Doug, this sounds like quite a challenging opportunity, and I certainly appreciate the vote of confidence, as well as the increase. But initially, wouldn't it be a better idea to make this a VP position?" It sounded like there were enough challenges without raising expectations and creating tension with Roger. Maggie might have an ego, but it was fed by accomplishment, not by titles or organizational power.

"No, Maggie, I want there to be no question. You're here to make some changes, and you'll have my full backing."

For the rest of the hour, they sorted out the details. As they wrapped up, Maggie asked what J. Randolph Barrister III thought about the situation. "I have the feeling Randy might not be as gung ho as you are on this move."

"You don't miss much, do you?" he said. "Since you asked, I'll be straight with you. Randy isn't a supporter of this change, and I already told you there are members of his team pushing to cut heads. He thinks Dynamic is in a mature market and may not be able to recoup its R&D investment. He's willing to let me try this, but I can't guarantee you that he won't pull the plug if he doesn't see improvements soon."

"Well, I appreciate your candor," she responded, not sure that she knew what she was getting herself into. But, after a moment, she looked him in the eye, offered her handshake and said, "Alright, Doug, I'll do it. I'm still not sure I know anything about new products, but I'll give it my best."

"I'm counting on it," said Doug as he walked her to the door.

Counting down

A few days later, Maggie finished her part in the TerraGra-fix transition. After all of the congratulations, well wishes, and farewells, she packed her belongings and headed home. It felt strange knowing that she wouldn't be coming back, but she was also looking forward to the challenge ahead at Dynamic Fluid Technologies. It had been a long few months, though, and before taking on a new challenge, she needed to reconnect with her family and recharge her batteries. Since her job change coincided with a school break, she and Jeff had decided to spend a week's vacation at a cabin in the North woods, enjoying some unplugged living and time in the great outdoors. It was something they hadn't been able to fit in often enough during the past few years, and she suspected they might not be able to do again for some time.

The Story

Part II – A Different Take on the Goal

Butterflies

Week 1 - After a relaxing week of vacation, at least as relaxing as keeping 9-year-old twins from finding new ways to torment their adolescent sister can be, it was back to the reality of home life and business responsibilities. As she drove to work for her first day at Dynamic Fluid Technologies, Maggie was surprised to find that she had butterflies. It reminded her of her first day at a new school.

Maggie arrived at the main DFT facility at 7:15, after the plant's shift change, but well before most of the staff was due. The fact that Maggie had worked at TerraGrafix made her employment check in relatively easy. She picked up her security ID which was waiting at the reception desk, noting that she had never liked the photo they had pulled from her file. Since Maggie was the new GM, the security officer assumed they could skip the review of the facility's safety rules and emergency procedures, but Maggie insisted that everyone had to go through them. "Even Mr. Barrister," she laughed, as the guard rolled her eyes, indicating that would never happen. Afterwards, they grabbed a cart and unloaded several boxes out of Maggie's car and wheeled them into her new office.

Located on the third floor, the office was far nicer than Maggie had expected. She certainly hoped they hadn't relocated Roger. If they had asked her, she would have told them to let him have the nicer office. She didn't need to throw gasoline on what she already expected to be a volatile relationship.

She spent the next few minutes putting away her personal effects and placing family pictures where she could see them. Before leaving on vacation, she had contacted Nancy Grimes, her new assistant, and given her a list of the operating reports and resource books she would need. Nancy had already efficiently arranged all of the reports on her desk. There was also a selection of books on innovation management, which Maggie spread out on the credenza behind her desk. She made no pretense about it—she didn't have any experience with new products and to try to hide that would be foolish. Maggie was here strictly because of her record of accomplishment in successfully running operations and her experience with improving processes. But she also knew that wouldn't be enough.

Just then, she heard a knock on the frame of her open door. "Hi, I'm Nancy. You must be Maggie. I wanted to welcome you and make sure that you have everything you need." She had an effusive smile and her white hair gave her a slightly grandmotherly appearance. Maggie had a feeling she could trust Nancy to look out for her.

Maggie had set an aggressive agenda for herself. Over the next week, she wanted to meet with most of the management team at DFT. Nancy gave Maggie a list of the meetings that she had been able to schedule. Looking at the list, she saw it was going to take a little longer than she had hoped.

"Sorry I couldn't get all of your meetings scheduled as soon as you wanted, but people around here stay pretty busy. What is it they say? Idle hands are the devil's playground or something like that."

"Well, you've given me plenty here to stay busy myself," Maggie smiled while gesturing at the reports.

Kick-off

Maggie started by reading over some of the reports that Nancy had pulled together, and before she knew it, her 9:00 reminder sounded, alerting her it was time for her kick-off session with Doug Stanton and all of her new reports. Included in the group were the directors for Product Management, Research & Development, Sales, Supply Chain, Finance, and Human Resources.

Even though he was visiting China, Doug had been gracious enough to introduce Maggie at the beginning of the meeting by video conference. "Good morning everyone, or afternoon or evening–whatever it is there," he joked. "I'm afraid the customers I've been visiting have a penchant for long dinners and strong drink. You'll have to forgive me if I seem a little sleepy." He explained that while DFT was performing well at the time being, his role was to look ahead. He was quite concerned with the new product pipeline and what wasn't coming out of it. He wasn't pulling any punches there. That's why Maggie was joining from TerraGrafix as the new EVP and GM: to bring her strong, continuous improvement skills to bear on the situation. Wrapping up, he said, "So, this morning, I want you to welcome Maggie to DFT. I know that all of you will give her your full support in continuing the strong operational performance that you're known for and in also delivering the strong, new product results that I know you are capable of." With that, he made a few final remarks and turned the meeting over to Maggie.

"Thanks for the kind introduction, Doug. I really appreciate you taking the time to join us from Guangzhou this morning, or evening, or whatever it is for you." After Doug signed off, Maggie spent a few minutes describing her expectations, her management style, and her immediate objectives. She hoped that would allay most fears about wholesale changes. They spent the rest of the meeting with each manager giving a brief overview of their responsibilities and top challenges.

Finding motivation

Back in her office after the kick-off session, Maggie continued to work through her stack of reports. She was particularly concerned with the new product pipeline report. Doug had not been exaggerating about the delays. She was almost overwhelmed by the sheer number of projects they were trying to run and couldn't make any sense out of the way they'd been prioritized. And if this report was any indication, most of the projects didn't have very clear objectives, either. Her first inclination, after reviewing the new product pipeline, was to take drastic action, but she knew that to keep people on board, she would have to proceed carefully.

Maggie's only one-on-one meeting that day was at 1:00 with Dr. Manoj Gupta, the head of R&D for DFT. She was glad to be speaking with him first. She would have been surprised if Doug's comments this morning hadn't made him uncomfortable—especially after reviewing the pipeline.

"Please, call me Manny," he said when they met. "I hope you don't mind if we walk while we talk," he said. "I'd like to show you the R&D facilities and introduce you to some of my team."

Maggie hadn't spent much time around the R&D folks while at TerraGrafix and wasn't sure what to expect from someone with a Ph.D. in fluid dynamics. She was delighted to find Manny very likeable and engaging. He was very early in his career, at least five years younger than Maggie, but his prematurely white hair made him appear quite distinguished.

As they talked, Maggie felt like she struck a raw nerve when she asked, "Manny, can you tell me a little bit more about the delays we're seeing in product development?"

"Maggie, my team is absolutely stretched to its limit. I've worked very hard for the last year and a half, since coming to DFT. We try to get to everything that Marketing and Sales want, but there are only so many hours in a day! To get any more done, we would need more people," he said flatly. Maggie would have expected anger or frustration, but if anything, he expressed resignation.

Having reviewed his file before the meeting, Maggie knew that Manny had come from a competitor to replace the previous R&D manager at DFT, who had lasted just two years. She wondered if there was a pattern there. It was an attractive promotion, giving Manny an opportunity that probably wouldn't have been offered anytime soon at his old company. Now, it looked like it was getting the best of him. That surprised her, given his previous accomplishments and background. Something wasn't adding up.

Glancing around to make sure they were alone, Maggie said, "Manny, I want you to know that I fully expect you to continue in your role. But, I hope you're open to change because a big part of my job is helping you identify where the problems are and resolving them."

As the R&D lead, Manny could have easily viewed the process of change as a threat. While he wasn't actually admitting that there were any problems, he almost seemed relieved to talk about it. She realized the poor guy probably suspected she was there to drop the axe on him, but she also suspected that many other departments were contributing to the problem. Of course, Maggie would have to make sure that he wasn't just cooperating for the sake of appearances.

They continued to tour the facilities and talk about the issues the team was facing. As they finished the discussion, he said, "Maggie, I assure you I am committed to making this work."

She hoped so, because she was going to need all the help she could get.

One down...

"Congratulations on your first day as General Manager, hon. We knew you'd get there someday," was Jeff's toast at dinner that evening. "And so quickly, too. Kids, aren't you proud of Mom?"

"Hooray for Mom," the twins chimed in while clinking milk glasses.

Even Sophia asked, "So, does that mean you're in charge of everything? Isn't that kind of scary?"

"No, dear, your Mom is a great leader. There's nothing for her to be afraid of," said Jeff. Maggie just hoped he was right.

After loading the dishwasher, the kids started on homework and Maggie and Jeff sat down in the family room to watch a favorite home improvement show that they had recorded. "So, tell me," he said, "What are your first impressions?"

"Well, it's too early to tell, but people seem to be awfully busy."

"Busy is good. Isn't it?" he asked.

"Sure. What do they say? *Idle hands are the devils playground*," she said, thinking of Nancy's earlier comment.

Enlightened & encouraged

The next morning, Maggie met with Danielle Espinosa the head of marketing. Petite, but athletic, Danielle exuded energy. She and her team of four product managers were responsible for deciding which products should be in the line and for pricing and positioning them in the market. Maggie recalled that Danielle had studied engineering and had earned her MBA in marketing.

When Maggie and Danielle sat down, Danielle jumped right in, "Any ideas on how we're going to get some of the high priority programs through the R&D group? I can't see how we'll make our numbers in the coming few years if we don't."

"I'd love to hear what you think the priorities should be, but let's not get ahead of ourselves," said Maggie. Then, she redirected the conversation so they could find out a little bit about each other. She also wanted to make it very clear that her approach wasn't to point fingers or assess blame, but to get the whole system of new product innovation working more effectively.

Maggie continued, "Danielle, what's the biggest issue your group is facing from a new product development perspective?"

"It's definitely time to market. The new W-3000 product we launched last month was almost a year late."

Maggie had already reviewed each project and had quizzed Manny about them, so she was ready for this one. "But I understand that some of the customer requirements were changed during design."

"Sure, but when competitors beat you to market, you sometimes have to add features. And always being late means that competitors frequently beat you to market!"

"Hmm," said Maggie knowing that Marketing couldn't be entirely blameless in this situation, but thought it best to let her vent.

"And the process just seems to delay everything. We're constantly waiting for gate reviews and the like." She was referring to the management reviews between steps in the stage-gate development process.[4] "I'm all for process, but some folks seem to want to use it as an excuse. There's a gate review coming up. Can't do any real work..."

All in all, it was an enlightening meeting. Danielle certainly had some issues with the way things were operating, but for the most part, it seemed she had a positive attitude and wanted to make a difference. If only all of Maggie's meetings that week could have been as encouraging.

Readers can download additional
information on the Speed-Pass™
innovation process at
www.GuidedInnovation.com/speedpass

Thinly veiled facade

She had been dreading it, but on Thursday morning, Maggie arrived punctually for her meeting with Roger Huntley, whose office was located at the opposite corner of the building. Obviously, Roger couldn't have been happy about being sidelined. And she wasn't exactly happy about him staying on where he could easily become a distraction. Well, he had a reputation of being great with customers. She could only hope that it would work out.

Roger was on the phone, but he motioned for her to come in and then raised his index finger as if to say, "This won't be a minute." From the sound of the conversation, Maggie thought that couldn't have been further from the truth.

"Listen here, Harry. We've got to close this if we're going to make the quarter... Let's worry about that next quarter...."

As she waited, Maggie couldn't help but notice all of the pictures displayed around his office. There were a few of family, but most of Smilin' Roger hunting, fishing, or playing golf with customers. She also recognized Randolph Barrister III smoking cigars and enjoying the outdoors with Roger in a number of the pictures. This guy was connected all the way to the top. She realized now that he'd probably made a lot of money for Randy over the years. Doug was really putting it on the line to make the change he had.

After five more minutes, Maggie motioned that she would come back another time, but Roger put up his hand and finally made a real effort to end the call.

When he finally hung up, he reached across the oversized desk and shook Maggie's hand vigorously. "Maggie, what a pleasure it is to have you join our team. Your reputation at TerraGrafix has earned you a lot of respect around here," his booming voice was only slightly softened by his stately, southern accent. He had come up through the sales side of the DFT business and definitely knew how to turn on the charm when he wanted to.

They spent the requisite time getting to know each other, but when they got to Maggie's agenda, Roger's demeanor took on a slight edge. He got straight to his issue. "Maggie, I've got to tell you that the new product delays have become intolerable," he said while pounding his hand on the table. It was almost like he was driving a nail. "I've given up on the ability of those boys down in R&D to be able to keep up with what customers want. Good luck making any headway with that bunch." He went on to explain that there were several projects running over a year late and that nothing big was coming out of the group.

While Maggie couldn't be sure that Roger was the one pushing to cut R&D headcount, she knew he wouldn't have opposed it. He was looking at retiring in a few years, and he probably figured that he could cut R&D by 30-50%, maximize his short-term bonus and build his retirement account—all to increase short-term cash flow and keep Randy happy. Unfortunately, that would also have left an empty new products pipeline and doomed DFT to either divestiture or the slow death of continual reengineering. Maggie was determined to make sure that never happened.

As they wrapped up the session, Roger walked her to the door, "Now, Maggie, you be sure and let me know how I can be of help."

While Roger couldn't have appeared any more welcoming, the encounter left Maggie with an uneasy feeling. She could only hope that he would recognize that her success would add to his, but she knew the road ahead would be bumpy. Below that gentlemanly southern façade, lurked someone with his own agenda.

Coming into focus

As Maggie spent the following days meeting people across the company, she was beginning to see a much more complicated picture emerging. It bothered her that she couldn't yet put her finger on the underlying problem. The R&D group in particular had some impressive, but very frustrated, people. "Innovation takes time. You can't expect us to work any faster," was the universal chorus. Roger may have wanted to fire most of them, but she got the feeling the better ones were already looking elsewhere.

Clearly, the group put in long hours. She also noticed that they were always busy—having another meeting to attend, experiment to run, or prototype to test. As everyone said, they were supposed to be busy—weren't they?

That Friday evening, Jeff and Maggie decided to have a family night out and treated the kids to their favorite deep-dish pizza. Maggie ordered a salad with light dressing on the side, but couldn't resist having a small slice of pizza, too. Oh well, she figured, she'd just run an extra mile or so tomorrow morning. When they got back home, Jeff offered to "entertain" the kids for a while so Maggie could catch up on some of her new responsibilities. Without saying it, she knew he was offering so that the weekend wouldn't be a complete washout.

After responding to a few emails needing her immediate attention, she reviewed a few of the innovation books that she'd brought home. Two hours later, she still didn't feel like she was making any progress and decided to call it a night. She wondered if she would ever be able to make any sense of the dizzying array of concepts: Customer-focused innovation, value migration, voice of customer, disruptive technologies, game-changing innovation, open innovation, outcome-driven innovation, profit pools, blue oceans, green oceans, pink oceans... It was all a blur. She felt like such a novice. While she didn't believe in shortcuts, Maggie knew she was going to have to find a framework—one that would help her make sense of all of this.

Serendipitous benefit

Week 2 - It was Thursday evening in her second week at DFT, and Maggie was driving across town to attend a fundraiser slated to benefit the area's technical college. Normally, she didn't make time for these types of events, but since Doug was still out of the country, he had mentioned that she might gain some good "developmental experience" by representing the Barrister companies in his place.

After arriving at the local art museum, where the fundraiser was being held, Maggie presented her invitation and then made her way to the bar. Pleased to find that the line wasn't too long, she intended to make a single glass of Pinot Grigio last through the entire event. Of course, driving home was always a consideration, but tonight the most important issue was time. She didn't plan to stay late, hoping to make it home before the twin's bedtime. She already felt guilty about not being there to help them with their homework and didn't relish the idea of being sleepy when she walked in the door.

Admittedly, Maggie hadn't had time to visit the museum since it had opened four years before, so this was a good opportunity for her to peruse its displays. She marveled at the generosity of the local manufacturers who had donated many of the nicer pieces. Even now, while they were falling on tough times, a respectable number of them turned out to support the technical college that had been a great source of talent as they had built their businesses.

After bumping into a friend who was still with TerraGrafix, the two spent the next hour wandering through the exhibits and catching up, stopping intermittently to chat with local dignitaries and faculty from the college. They were discussing the tough local economy with the general manager of a local stamping plant when Maggie glimpsed a familiar face, standing just a few exhibits over. When a break in the conversation allowed, she excused herself and walked over to greet him.

Professor Isaac Yulinski was in his early seventies and a little on the short side, but exceptionally fit. With a fringe of clipped white hair, a goatee, and little round glasses, he could have easily been underestimated, but his intense green eyes belied his natural curiosity and energy.

As he admired a beautiful red and black samurai sword, Maggie interrupted his gaze. "Professor Y, I hope you're not planning to add that to your collection."

"No, swordplay has never been my strength," he replied. Turning around, a warm smile conveyed his delight to see her. "Maggie, my dear, how have you been? I haven't seen you since we completed the project with your team at TerraGrafix."

She could tell that he was pleased to hear her use his nickname. Of course, it was easier to remember and pronounce than his given name, but she also knew he enjoyed the nickname that his students had given it to him because his favorite teaching tool was encouraging them to always ask 'why.'

"I'm doing well, thanks. You probably know that I left TerraGrafix after the sale." She went on to explain the events of the last few months.

When she finished, he asked, "I'm curious, Maggie, why did you decide to take on such a different assignment?"

"I'm beginning to ask myself that," she laughed. "But you know that I love to try and accomplish the impossible."

"Indeed, but is this new role really so impossible?"

"Well, in all seriousness, there's a lot to learn about new product development. There's more written on the subject than I could have ever imagined," she said. "It's almost impossible to know where to start."

"Hmmm...How is the organization responding to the change?"

Well, I'm certainly running into the resistance you would expect. You know, the stuff about not being able to treat innovation like an assembly line."

"Yes, all acts of creation include an element of art, and ever since the industrial revolution, the argument has been one of art vs. science. Of course, Henry Ford and others manufacturers of his era faced the same argument almost 100 years ago."

"That's right—it was the artisan or the craftsman era, wasn't it?" she asked.

"Yes, if companies had accepted that as fact, we would still be buying our shoes from the cobbler. You still can, of course, but they'll set you back a lot more than a pair of Allen Edmonds," he winked as he raised the toe of his shoe.

"Maybe even Prada," she laughed.

"You do have some sort of staged development process, don't you?" he continued.

"Sure, we use a stage-gate process, like anyone else our size. But the technical folks just think it's our way of keeping track of them and claim that it only slows them down. They may be right. Honestly, I think we spend too much time working around it."

"Why?" he asked, living up to his reputation.

"Why does it slow them down? Or why do we work around it?"

"Maybe both. It might be the same reason."

"Are you saying that we need to change our process?"

"Oh, my goodness," he said, looking at his watch. "I'm afraid that I'm going to have to get going if I'm going to make it to the airport in time. I only intended to make a quick appearance this evening and then be off to spend some time with my grandchildren in Portland."

"Would it be alright if I walked with you to your car?"

"Certainly, my dear."

As they made their way toward the exit, Maggie asked, "So about our new product process?"

"Yes. Well, I mean the answer is no. I'm not suggesting that you change your process—unless, of course, it's limiting your throughput."

As he looked back, he caught the quizzical look on her face.

Without breaking stride, he said, "Do you remember the work your team did to de-bottleneck the assembly process at TerraGrafix?"

"Sure. I wasn't involved directly. In fact, I still have most of the resource materials you provided. But I don't understand what that has to do with new products." She was hurrying to keep up as they headed down the wide outside stairs.

"While new product innovation isn't my area of expertise, either, it seems to me that you need a similar approach to simplify improvement of such a complex subject. Maggie, yours is an intriguing problem. Tell me, what do you think is the goal of innovation?" he asked as they headed across the parking lot.

"Hmm...I suppose it must be to achieve some kind of game changing or disruptive advantage." Even as she said it, she knew that it didn't feel right. "At least, that's how the corporate folks talk about it."

"Yes, I think their pay grade must be determined by the number of buzzwords they can work into a sentence or how complicated they can make a simple idea," he chided.

They arrived at his car, and he opened the door. With one foot in the car, he leaned toward her to offer his hand and said, "I'm afraid it's time to say goodbye."

"Just when it was really getting good!"

"Indeed. But I'm going to leave you to answer that question, Maggie. What is the goal of innovation? And how would you measure improvement? Think about it. I know the answer will come to you." The wheels were always turning with the Professor. "In fact, the more I think about it, the more certain I am that elements of the approach that you learned in your manufacturing operation apply to your innovation problem."

With that, they shook hands, Maggie wished him a safe trip, and he was off.

Maggie looked at her watch and realized that she, too, had lost track of time. She headed to her car, knowing she wouldn't make it home in time to help with homework. That was a good thing, because she now had homework of her own.

Late night reflections

Maggie made it home just in time to help tuck the twins in. As they walked down the hall to say goodnight to Sophia, Jeff asked how her evening had gone.

"You remember Professor Y, don't you?"

"How could I forget," he said holding up a slightly wayward little finger.

"Oh...that's right." How could she have forgotten? Her six–foot four-inch bear of a husband had broken his pinky in a Tae-Kwon-Do class led by the diminutive senior who happened to be a third-degree black belt. He'd worn a splint for two weeks afterwards.

"You'll be happy to know he's given up sparring," she teased while kissing the finger.

"Too many complaints?" he laughed.

"Probably. Anyway, we had a very interesting conversation about getting more from product development."

"A conversation? With Professor Y?"

"Okay, more like him asking questions and me not having the answers, but he gave me some things to think about. In fact, I'm going to go dig out a few things from the study."

He knew what that meant. "Alright, I'll see you in the morning then."

"No, I'll just be a while."

"I've heard that before," he said gently patting her behind as he headed down the hall before she could protest further.

At least he understands me, she thought as she hurried into the study. Study might have been an understatement. It was actually one of the largest rooms in the old house. It was even equipped with 12-foot tall floor to ceiling shelves and a rolling ladder to reach the upper sections. Unfortunately, in the house's recent history, someone had painted the room's woodwork. She shuddered to think what a mess it was going to be to restore the room to its original mahogany finish. But she was glad there was so much shelf space because she was able to quickly locate the cardboard file box of books and notes she had been anxiously looking for.

The work they had done at TerraGrafix used a focusing approach called *Theory of Constraints* or TOC. She remembered now that the Professor had always said it should have been called the *Law of Constraints* instead, since it was clearly more than theory. *'Just ask the thousands of companies across the globe that had used it to drive manufacturing improvement,'* he had remarked.

As she sorted through the resource materials, she also recalled that much of the literature on TOC was written in business novel format. She spread some of the material out on the large library table that had come with the house. Eli Goldratt's book, *"The Goal,"* was the seminal work on TOC, and it wasn't long before she found the part where Alex Rogo, the main character, began his quest to discover the goal of his business.

There it was. *"The goal of any manufacturing organization is to make money."* She already knew that, and she could see a connection to new products, but she knew there must be more. She dug in further, and about thirty minutes later, she found the next clue.

"The goal is to make more money now and in the future," she read aloud, causing Sweetie to open an eye to see what had roused her master. This goal related to a manufacturing organization, but she could clearly picture how it could be applied to innovation. The goal of new product development wasn't just to make money. It was very specifically to support the organizational goal by ensuring that the company would make more money in the future. She'd have to think about what that meant to her task of improving new products at Dynamic Fluid Technologies, but she knew there was something here.

Playing hooky

The next morning, Maggie surprised everyone at the breakfast table by declaring that she was going to work at home for the day.

"Are you sick, Mom?" asked Luke, as his brother, Leo, nodded for emphasis. Even Sophia took out her ear buds to hear more.

"No, of course not, I have some studying to do, and this is the best place to do it." Maggie responded.

"Like school?" This time Leo did the talking and Luke did the nodding.

"In a way, I guess it is," she said.

After they got the kids all packed and in the car so Jeff could take them to school, Maggie left a quick message for her assistant, Nancy, asking her to reschedule a few meetings. Then she settled into the most comfortable chair in the study to dig through the rest of the material on TOC. Sweetie seemed glad to have the company, and after getting her ears rubbed, quickly made herself comfortable next to Maggie's chair.

With all of her other responsibilities at DFT, she didn't have time to devote to this research. Staying home on a Friday also made Maggie feel like she was playing hooky, but she sensed that what she was working on was going to be very important in the coming months.

By the end of the day, she still had a lot to cover, but she had distilled some of the key points and laid them out on the table in front of her.

Coming from a manufacturing background, TOC had always seemed like common sense to Maggie. But, as the saying goes, 'Common sense isn't necessarily common practice.' She couldn't remember who said it, but they were right. Of course, she had not yet seen it applied to innovation, either. Maggie couldn't wait to ask some more questions at DFT and see if some of these concepts could help them to focus in on their key issues.

TOC - THE BASICS

- The goal of any company is to make more money now and in the future. (Innovation and marketing working together)
- Metrics - Real improvement vs. goal achievement is measured by the change (Δ) in T, I, &OE:

  ΔT: Increasing Throughput - cash flow over and above truly variable costs

  ΔI: Reducing Investment - cash tied up in the business, including working capital items such as inventory and accounts receivable

  ΔOE Reducing Operating Expenses - costs of running the business, not including raw materials

- Note that Sales Throughput is preferred metric - not profit which depends on allocations that can create distortions.
- The key to improving performance vs. the goal in any process is the step with the lowest capacity, since that is what limits the global systems throughput.
- Constraints are the physical limitations and policies that prevent the system from selling more.
  (Bottleneck and constraint are often used interchangeably - pay attention to context)
- Time saved at the bottleneck can rapidly increase throughput.
- More efficient use of a non-capacity constrained resource does not increase throughput- only gives illusion of efficiency.
- Traditional improvement provides less leverage since it encourages improving the entire process, including non-constrained resources.
- TOC provides high-leverage improvements – Almost 100% of the improvement comes from work done at the constraint.
- Any Improvement effort must answer four questions:
  1. What is our goal?
  2. What to change - the constraints on our bottleneck?
  3. What to change to - the way we operate the bottleneck?
  4. How to cause the change - involve people that will be affected?
- People resist change if they feel it threatens their security - involve them in using TOC to answer these questions and they will take ownership for the improvement

- There are five focusing steps to achieve high leverage improvement:
  1. Identify the constraint
     - Where is uncompleted work stacking up?
     - Where are downstream groups waiting idle?
     - Bottleneck can be constrained by physical resources or by our own operating policies (self-imposed constraints)
     - Market can also be a constraint (Need to think about what this means in relation to innovation)
  2. Exploit - how will you operate the bottleneck to get all of its capacity?
     (Adding capacity requires investment- deferred to step 4)
  3. Subordinate - how will the rest of the operation help in exploiting the constraint?
     (Before each step, we should ask if the bottleneck has been broken. If it has, we should proceed to Step 5)
  4. Elevate - how can you add capacity to break the constraint? (This step usually requires investment)
  5. Start Again - What is the next constraint?
     Prevent inertia from becoming the next constraint
     Evaluate policies put in place to protect the old bottleneck

The Story

Part III – Constructing the Framework

 **Step 1 – Identify**

Seeing the treadmill

Week 3 – After a whirlwind of a weekend spent between the twins' soccer matches and Sofia's brown belt test in Tae-Kwon-Do, Maggie was anxious to talk with her team about the things she had learned. But the first thing on her agenda for Monday morning was to meet Harvey Watts, the head of Human Resources for DFT. HR was located on the ground floor, two floors below Maggie's office, but she decided to take the stairs. When she arrived at Harvey's office, he was sitting at a small conference table and talking on the phone. She was a few minutes early, but he still motioned for her to come in.

"Sorry, but I'm going to have to call you back, Phil," he said, ending his call. After hanging up, he turned to her. "Hi, Maggie. It's a pleasure to finally get to talk with you one-on-one. I'm sorry I haven't been available since your intro session."

"No problem, Harvey. I'm glad we're getting the chance now."

They spent the next 20 minutes talking about how things had been going since Maggie's arrival, what her objectives were, and what she had learned about some of the problems. This prompted Harvey to ask, "If you had to boil it down, what do you see as the biggest problem so far?"

"As you know, we need a better return on our new product investment, but I'm finding that the R&D folks already feel stretched too far—almost to the breaking point. Of course, their answer is to add more resources. I guess you can't blame them, though."

"No, you can't," he agreed. "But we know that won't fly."

"Nor should it. Of course, that answer isn't any different from the one I always heard when I headed up manufacturing for TerraGrafix. But there's always hidden capacity there."

She paused to reflect. "Harvey, are you seeing requests for transfers to our other divisions?"

"Why do you ask?' he asked guardedly.

"Look, I don't want specifics, but do we have an unusually high number of R&D people applying for jobs in other divisions?"

"Not just in R&D," he admitted.

"Honestly, I'm not surprised. Tensions are high, and we have some pretty frustrated people."

"Do you think they might benefit from training in time or project management? You know, to help them get more done."

Maggie considered whether that was really what was constraining the group and decided it didn't sound right. "Harvey, I appreciate your suggestion and eventually that might be something that would help. Right now, though, it seems to me they might view it as an extra brick on the heavy load they're already feeling. That's probably not ideal for an already stressed out group."

"I understand. You wouldn't want to give the impression that you're just here to get them running faster on the same treadmill."

"Wow, I like that metaphor. Instead of expecting everyone to run faster, we need to understand if there's a root problem we're not seeing. If there is, we can figure out how to make sustainable improvements."

They spent the rest of the meeting getting to know each other. After 20 minutes, Harvey had to head off to another meeting. "Well, let me know how I can help. I'll be happy to pitch in if there's any kind of training or job redesign that would help," he said as he hurried off.

As Maggie walked back to her office, she couldn't get the treadmill image out of her head. She could see the new product group all pounding away on treadmills, with Manny out in front. Busy, busy, busy, but going nowhere. It wasn't just the R&D group, either. It wasn't working, and she knew there was a different approach. They just had to find it.

No time like the present

Upon returning to her office, Maggie noticed a new pink message slip on her desk. *Damn,* she almost said but caught herself. The Professor had called, and Nancy had taken the message: *"Wondering if you had answered that question yet? Staying longer than planned. Only available by phone till 8:30 AM."* She'd missed her chance—that meant she'd have to solo for a while longer.

Maggie decided that there was no time like the present. Manny had been in his office down the hall when she'd passed by it earlier. She popped her head in and said, "Good morning, Manny. Do you have a few minutes?"

"Sure. I have a meeting in about 45 minutes, but I don't have anything pressing until then."

"Great, let me grab Danielle and we can talk in my office."

After comparing weekends, Maggie focused on the business at hand.

"Can you tell me what you think the goal of innovation should be?"

"Absolutely, to develop game-changing new technologies." replied Manny, as Danielle nodded in agreement.

Maggie shook her head, realizing they were just saying what they thought a corporate exec would want to hear. "And why is that so important?" she continued.

"To give us a competitive edge in the marketplace," said Manny.

That sounds about right, she thought, before using the Professor's tactic. "But why do we want that competitive advantage?"

Without really answering the question, they looked at her skeptically. "Maggie, that's just Business 101."

"I know, I know, but please give me the benefit of the doubt for a few minutes," she laughed trying to lighten them up. "I'm just trying to understand some things, and it's helpful to talk them out. It helps to make everyone's assumptions explicit."

Their skepticism appeared to lighten—if only a little.

"Humor me. I promise this isn't a test," she said. "So, why do we want to do that—the competitive advantage part?"

"So we can sell more than the competition." Danielle was on the right track, and they were both becoming curious about where she was headed with this line of questioning.

"Now, bear with me just a little while longer. Why do we want to sell more than the competition?"

"Survival?" asked Manny.

"How does that help us survive?"

"By making more money," Danielle volunteered.

That wasn't so painful, was it? she thought. "Then would you agree that the goal of our new product innovation efforts are to ensure that DFT makes more money in the future?"

Manny jumped in. "Well, Sales would probably say we're here to make the customer happy. Human Resources might argue that our goal needs to include a fully engaged workforce. But, ultimately, they all support that same goal. Yeah, making more money in the future sounds right to me."

"But that's pretty obvious, isn't it?" Danielle asked.

"Maybe so," said Maggie. "But a few minutes ago, the goal was to create game-changers. By being more precise, we've created something we can measure—a standard that tells us whether improvements are real—whether they result in us making more money now or in the future."

"Okay, I can see that," said Danielle.

"Good. What we've done here is identify our goal. That's an important foundation for an approach that I believe is going to help us drive the innovation improvements we need. It's called *Theory of Constraints* or TOC. It was very effective in doubling production capacity in the TerraGrafix division without adding a single resource."

"A manufacturing approach?"

"A systematic framework for continuous improvement. And while it did get its start in manufacturing, it's being used across companies now because it helps identify the place where small changes have a big effect."

Danielle nodded, but Manny crossed his arms and looked uncomfortable again. "I still don't know how my group can work any harder."

"Maybe it's not about working harder." This time it was Danielle.

"That's right. Manny, I'm not saying that your people are the problem. It might be the way we're asking them to operate that's the problem. But, in either case, you both know that Randy and Doug expect more growth out of our investment. That can't be ignored."

That statement hit home, and they nodded in agreement.

Maggie walked up to the whiteboard. "You know it might be helpful if we sketched out the workflow, including all the steps from the time an idea is hatched until the product is launched."

"That's easy," Danielle said. "We did that already when we implemented stage-gate."

Danielle took Maggie's place at the whiteboard. In five minutes and with a little help from Manny and a few clarifying questions from Maggie, they had it all mapped out.

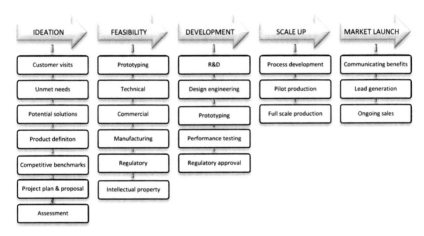

"Alright then, going back to our goal, what's limiting our ability to make even more money with new products?"

"Not enough people." Manny's instant reply didn't surprise Maggie—in fact, she'd expected it.

"In which department?"

"All of them."

"Sure, every step has bottlenecks," agreed Maggie. "But one has to be the biggest problem—the one that's constraining the process, maybe even the overall system."

"Design Engineering," he replied.

"Why do you say that?"

"Well, they're the furthest behind and working the longest hours. Plus, the detail drafting group is always complaining that they're waiting for design specs from them so they can take it to the next stage." Manny had just put his finger on two of the key symptoms to help diagnose a bottleneck. "So, that's where we need to add resources," Manny stated, not giving up easily.

"Manny, one of the things you'll learn as part of TOC is that we should look for the highest leverage changes first—we search for areas where small, inexpensive changes can make a big difference."

"Like Judo," observed Danielle. While she might not have looked the part, Danielle had a black belt in mixed martial arts. "Use the other guy's momentum to drop him," she explained, making a flipping motion.

"Yeah, kind of like that," said Maggie, raising her hands in mock defense. "But my point is that we don't want to add the fixed costs of more equipment or staff until we are fully exploiting our existing resources."

"You mean using 100% of their hours?" asked Danielle.

"What's more important is to make sure that we get the most benefit possible out of the hours we do use—working smarter with our existing resources. In fact, we normally try to keep the utilization from going above 80-85%. The delays get exponentially worse as you get to 100% utilization, and that just ends up reducing overall system output.[5]

"Oops," said Manny checking his watch. "I'm going to be late for a meeting, which would be especially bad form since I called it. But, Maggie, I do want to finish this discussion. How about 3:00 this afternoon—same place?"

"Could we make it 4:00?" asked Danielle, looking at the calendar on her Blackberry.

"Okay by me," said Maggie, waiting for a nod from Manny. "In the meantime, here are some notes on the TOC approach that we used in manufacturing. Try to review them if you can. It'll help us take this discussion further."

Already, Maggie was beginning to wonder how she was going to break the culture of activity that had a grip on DFT and replace it with a culture of results. "I'm intrigued," Danielle said, looking up from her Blackberry. "Is TOC some kind of lean manufacturing approach?"

"Actually, it's more of an overall framework for focusing those improvement efforts. In fact, some companies have seen their Lean and six-sigma efforts deliver 15 times more results by using TOC as a focusing mechanism."[6]

"Wow, are there any books available on the subject?"

Maggie laughed as she walked to her desk. "More than a few. Here, you can make a copy of the reference list I started. They're not all written like traditional business books, but I have to warn you. They're addictive."

Readers can download Maggie's
reference list at
www.SimplifyingInnovation.com/extras

What's critical?

Later that afternoon, Maggie was immersed in the latest financial reports when her office phone rang. Thinking that it might be Jeff or one of the kids, Maggie grabbed it quickly. Instead, she was surprised to hear the Professor's voice, scratchy though it was, on the other end.

"Maggie, my dear." She smiled at his antiquated, but well-meaning, greeting. "Sorry I missed you this morning. This is probably the last chance we'll get to talk for a while. I'm afraid the reception here in the mountains is one bar at best."

She pictured him trekking up Mt. Hood. At his age, he'd still make it twice as fast as most people. "I'm glad you called. I've got so many questions for you." She spent the next five minutes outlining the steps they had taken so far.

"Sounds like you're making progress, Maggie. Now that you've identified the bottleneck, what do you plan to do next?"

"Well, we really need to understand what's constraining it.

"And what is that likely to be?"

"Doesn't that depend on the situation?" she asked.

"Yes, but what will many of the constraints have in common?" Maggie was stumped. She enjoyed the Professor, but he really made you work for it.

He prodded her further. "Peter Drucker offered a valuable clue when he said, *'Most of what we call management consists of making it difficult for people to get their work done.'*"

"Oh right—policy constraints. That makes sense. The most common constraints are our own management policies."

"And what else can you say about them?" he continued.

"They can be tightly held, but the return should be very high for very little investment—high leverage as they say."

"Exactly. Now there's one other thing I wanted to cover. It's critical..." and with that the line crackled and went silent. She waited a moment for him to call back before she redialed him, but it went right to voicemail. Apparently, he was out of range. *What's critical? What do I need to look out for?*

Breaking new ground

At 4:00 that afternoon, Manny and Danielle returned to Maggie's office. "Now, where were we?"

Manny jumped right in. "We were talking about the capacity of the engineering group."

"That's right. In the handout I gave you, you'll see that once we know what the constraint is, Step 2 is to decide how we are going to best use or fully exploit its available capacity."

"So, are you saying my engineers are wasting time?" Manny was getting defensive again.

She was glad that the Professor had jogged her memory about policy constraints. "Manny, there's wasted time in even the best of operations. But it's not necessarily their fault," she said motioning in a calming gesture. "If anything, as a company, we may be forcing them to work in a wasteful, less than ideal manner."

"As a company?" It was starting to sink in for Manny.

"Yes. They're called policy constraints. The constraints organizations unintentionally place on themselves with well-meant, but misdirected rules, policies, and practices."

"Sounds a little like the law of unintended consequences," observed Danielle.

"Yes, in a way it is. Let me share an example," said Maggie. "When I first started at TerraGrafix, we had a six-days-a-week operating policy in manufacturing. The production manager was religious about preventive maintenance, and the seventh day was when maintenance came in to do the PM's. When I asked what the capacity was, everyone said 10,000 units because they had made 6 days an implicit assumption. I asked how long PM's for the bottleneck took and found out that it took only an hour or so. So in reality, the capacity should have been based on 7 days less 1 hour, or 11,600 units. By moving to a 7-day operation with PM's scheduled around non-bottleneck idle time, we immediately boosted capacity by 16%. And that increased the bottom line even more since all of our capacity was already sold out."

Those results elicited a whistle from Danielle.

"Okay, but we can't work my group any more hours than we already are," said Manny.

"Agreed, and I don't think we're going to need to, either," Maggie replied.

Skepticism started to seep back into Manny's expression, "I'm still not sure where you're headed with this."

"That's because we're figuring it out together. Remember, TOC grew out of the manufacturing arena. Applying it to innovation is breaking new ground." Manny's interest was finally piquing.

"Alright, so how would you see us implementing this here at Dynamic?" he asked.

"Let's not get ahead of ourselves. How soon can you get your key project and product managers together for a half day?" asked Maggie. "If we're going to have any success at all with the changes, we're going to need them along on the journey."

"Why is that?" asked Danielle.

It had been a while since she'd thought about TOC's change process, so Maggie hesitated, jogging her memory for a few seconds. "Well, there are four questions to answer if we want to make real and lasting improvements: one, we must know what our goal is; two, we have to know what it is that needs to change—in other words, we must identify our innovation bottleneck and what's constraining it; three, we need to decide what to change to—what should we do differently to move closer to our ideal innovation capacity; and four, we have to determine the best way to cause that change."

"That's why we need a group session. People will resist change if there's any question whether it's in their best interests. We want folks to feel secure with any changes. So we need them to be involved in answering these questions and in determining the best way to make those changes. Hopefully then, they'll take ownership."

"Why can't we work it out ourselves and explain it to them later?" asked Danielle. "As busy as they are, I honestly don't think they'd complain."

"Danielle, by now I think you know that I'm all about getting it done, but it's more buy-in that we want, not less complaints. If we brainstorm and develop this by ourselves and then spring our masterpiece on them, the best we can hope for is compliance. They'll feel no ownership and might even resist in subtle ways. Worse yet, without the context they'd get from being involved in the process, they could view it as another clever re-engineering move to do away with more jobs over time—and more fear isn't what we need."

"I've seen what you're describing before," admitted Manny. "It's so-called passive-aggressive resistance. Without any involvement, some folks won't say a word at the beginning. Then as soon as we run into any little problem, they're the first to gleefully point fingers."

"Exactly," said Maggie.

"I hate to pull folks off of their tasks, since some of them are working in the bottleneck," he grinned. "But I can see it's important. How about next Monday morning? We can get together in place of my staff meeting."

With a few minor schedule adjustments, Maggie and Danielle agreed that would work. After everyone had left, Maggie added a few notes to the TOC for Innovation overview she was creating. She suspected it was going to get a lot of use.

Readers can download a complete set of
Maggie's TOC notes at
www.SimplifyingInnovation.com/extras

Step 2 – Exploit

Getting the ball rolling

Week 4 – One thing Maggie had noticed was that across the organization, DFT had suffered from what she called *Device Attention Deficit Syndrome* or DADS. It appeared that people didn't think there was anything wrong with stopping in mid-conversation to pick up their mobile phone to see who was calling or texting them. For Maggie, that sent the implicit message that if it was someone more important, they might take the call. Even worse, during meetings, they constantly checked email on their laptops or phones, even pausing to send replies–the false economy of multi-tasking, like heroin to an activity junkie.

To combat potential distractions like these in the team meeting, Maggie arranged for the use of an off-site conference room. She needed the group to focus solely on what they were doing this morning, so the meeting invitation made it clear that no one could bring laptops into the meeting room. To everyone's objection, she'd also placed a box at the door to hold everyone's phones and electronics—no one was exempt, not even Maggie. Parting with their communication devices wasn't easy, and Maggie was particularly surprised to see Danielle's reluctance to let go of her Blackberry.

When everyone was seated, Maggie quickly reviewed the purpose of the meeting and got the group's additions to the basic meeting guidelines. She needed a little bit of crisis to make sure they took this seriously, but not so big that it would shut them down. So she started by talking about the increasing delays in delivering new products and then set them against the market share gains that the competition had recently made. She also knew that the sale of TerraGrafix had resulted in an unspoken threat looming in the other divisions. So, she pointed out that Barrister wouldn't be investing in this exercise if the health and growth of DFT weren't important.

The scene set, she spent the next hour using the Socratic teaching method to take the group through the same logical exploration of goals and bottlenecks that she, Manny, and Danielle had covered. When she was done, a member of the team asked, "What if we have more than one bottleneck?"

"That's a possibility," she admitted. "But it's pretty unlikely. Some of the most complex systems are governed by only one bottleneck at a time. That's what makes this approach so powerful. Even highly complex systems tend to have only one constraint. That greatly simplifies matters and gives us a powerful leverage point for improvements."

After answering a few more questions, she said, "There you have it. Now you know as much as we do." She included Manny and Danielle by motioning in their direction, and then waited for the inevitable reply.

"But I thought you were supposed to be the experts," pointed out one of Manny's project managers.

"I know a few things about TOC, and I've used it to drive improvements in manufacturing," she replied. "As far as applying it to new product innovation, I'm afraid we're going to have to learn that together." She paused again to let that sink in. "Besides, even in a manufacturing environment, TOC isn't about experts telling you what you need to do. It's a set of tools they use to help you discover that for yourself."

His response, "Fair enough," was followed by the agreement of several other team members.

"Are there any other concerns?" Maggie asked, inviting them to air any doubts they might have right away.

"Well, I can definitely see how this TOC approach applies to manufacturing. In that environment, it's actually a pretty powerful concept. But innovation is not like an assembly line that you can just turn on and speed up," said one of the other project managers.

"Why is that?" Maggie asked her.

"Because sometimes somebody has to invent something," she replied while holding up a broken filter prototype that she'd pulled out of her bag. "And you can't schedule invention."

"Okay, I'll give you that," laughed Maggie. "But would you at least agree that invention, or solution development as we call it, is still a step in the process, even if it isn't completely predictable?"

"Hmmm..." that question invited further thought.

"And aren't there even tools for improving the ability to invent solutions?" Maggie pressed on.

"Well, I suppose there are tools like Lateral Thinking[7] or Theory of Inventive Problem Solving, so-called TRIZ,[8] that are meant to help in that regard. Is that what you're thinking—that we just need to learn to be more creative?"

Standing on the sideline, the look on Manny's face told Maggie he was waiting to see how she was going to dig herself out of this one. "That might be very helpful if the bottleneck was coming up with solutions. But it doesn't seem like that's our root problem."

"You know the problem I have with treating innovation like a manufacturing process is the project element." This came from Ed, one of Danielle's product managers.

"What do you mean by the project element, Ed?" asked Manny. Maggie appreciated that he wasn't making her carry the entire discussion.

"I mean that new products programs are managed as individual projects."

"But what about stage-gate?" asked Maggie. "Isn't that a process for managing the project?"

"I guess," he conceded. "But maybe that's what I'm struggling with. New Products are a hybrid of project and process. How does TOC apply then?"

"We're going to have to think about the implications for that," admitted Maggie. "Let's put that on the flip chart to make sure we don't forget it." Without the Professor here to back her up, she hoped she wasn't getting in over her head.

Building momentum

An oversized diagram of all of the steps for the new product process hung on the wall. Maggie pointed to it and said, "Okay, so everyone seems to agree that the bottleneck is in Design Engineering."

"Yeah, but I still don't like being called the bottleneck," complained one of the engineers.

"Yes, but that also means you get to be the hero because the bottleneck is the only place we can achieve real improvement," teased Danielle. Maggie realized she had done some studying since their earlier discussion.

Keeping things moving, Maggie approached the flip chart. "We could try to design the ideal process from a blank sheet of paper, but I think it's more practical to start by looking at our existing situation and see what's constraining our bottleneck."

"But I thought there was only one constraint that we needed to focus on."

"I'm glad you brought that up," said Maggie. "That's something I didn't understand for quite some time. The simplicity of this approach is that there is rarely more than one bottleneck acting at a time. However, many things can act to constrain that bottleneck." Then, while writing on the flip chart, she said, "They can be..."

Constraints Acting on the Bottleneck
- Policies
- Procedures
- Capabilities
- Availability of resources
- Quality of inputs
- Suitability of operating techniques

"Now, this might look like a complicated list, but the beauty of it is we don't have to look all over the organization—just at the bottleneck. Does that make sense to everyone?"

"I just want to make sure I understand this," interjected one of the product managers. "You're saying that if we see a way to improve the speed of our prototyping step, that we shouldn't do it? I thought we were always supposed to strive for improvement everywhere."

"Not always, but would speeding up prototyping allow us to push more through Design Engineering?"

"Well, no, but still..."

"Would it be an improvement then?"

"Hmmm...I not sure what you mean."

"Let's put it a different way. If it was your money at stake, would you invest in speeding up the prototyping group, knowing that the improvements wouldn't get you any more sales?"

"I guess not—at least not until it becomes the bottleneck itself."

"That's exactly the way we need to look at it," said Maggie. "The way Goldratt puts it is that '*An hour lost at a bottleneck is an hour out of the entire system. An hour saved at a non-bottleneck is a mirage.*' "[9]

Moving along, she said, "So, our next step is to find out what things are constraining our bottleneck. Any ideas?" For the next ten minutes, the possibilities, mixed with a fair share of venting, flew back and forth. Occasionally, Maggie referred to the constraint types to keep them on track, while moderating just enough to keep it from degenerating into a complete bitch session. It wasn't long before they'd narrowed it down to a reasonable list of issues:

DFT's Bottleneck Issues
- No remote access so work can be done at home
- No in-depth planning or planning time
- Too many reports to fill out
- Too many meetings
- Too many emails to read
- Not enough technician help
- Frequent switching of priorities
- Delays in requests for more information
- Changes in specifications or features after design starts
- Resources pulled away to work on sustaining projects
- Selected technology doesn't always work

At the end of the exercise, Maggie thanked them for their contributions and suggested they take a ten-minute break. Their phones stayed in the box—Maggie's way of making sure their break wasn't a minute longer than she'd allotted.

Evaporating clouds

When everyone had reassembled, Maggie decided to try Harvey's metaphor on them. "As you can see, there's a sizeable list of issues we could attack here, but addressing some of them now would just be asking you to run faster on the same treadmill. It still feels like we're tiptoeing around the root cause..."

She had their attention, so she paused and walked across the front of the room. Borrowing from the Professor's Lean technique of asking why to get to the root cause,[10] Maggie asked, "Why are there so many reports and meetings? Why aren't there enough technicians? Why do priorities frequently change?" Not getting any response, Maggie rephrased the question. "What drives the number of reports and meetings?"

"The number of projects," Danielle replied.

"Does that mean then that we're working on too many projects?" Now she had their interest. "How many of you are compensated by the number of projects that you work on?"

Manny looked uncomfortable again, like he wasn't quite sure where she was going with this.

Maggie continued. "None of us are paid by the project, are we?"

"Well, no..."

"Ultimately, what are we paid to do?"

"To get successful new products to market," said one of the project managers.

"That's right," said Maggie. "So, let me ask the question again. Could working on too many projects at the same time be constraining our ability to get successful new products to market?"

"Maggie, if we want to increase throughput, we need to do more projects," Danielle reminded her, while not-so-subtly defending her own wish list of new products.

"We clearly have too many projects," said Manny. "But, Danielle's right, corporate wants more output—not less."

"Manny, you've just put your finger on an important conflict," said Maggie, walking back to the whiteboard. "You know, there's a TOC thinking tool that might help us resolve this." Drawing a box as she spoke, she said, "Our goal is to make more money, so consistent with that, our new product objective is to increase new product sales throughput." Then just to the right she drew two more boxes, above and below, with arrows pointing back to the first box. "We are considering two different approaches for accomplishing that. Now, I've only used this approach once before, so you'll have to bear with me," she said, contemplating the diagram. After a few seconds of silence, she moved to the top box and wrote, *'Requirement 1: Launch greater number of new products.'* In the bottom box, she wrote, *'Requirement 2: Deliver new products faster'* "There, I think that's right." she said. She finished by creating a final set of boxes to the right, again above and below, feeding into the previous two. These she labeled. *'Prerequisite 1: Increase the number of active projects,'* and *'Prerequisite 2: Decrease the number of active projects.'* Stepping back to assess the logic of what she had written, she said, "It's pretty obvious these two prerequisites contradict each other." She finished by drawing a bent arrow to indicate the contradiction.

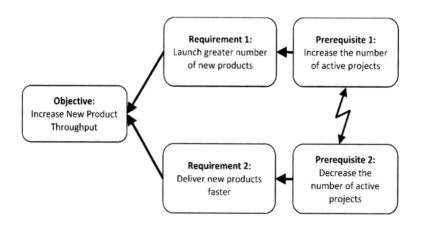

When she was done, she explained that TOC called this conflict resolution diagram the "Evaporating Clouds Method."[11,12] By looking at the underlying assumptions, incorrect approaches, or clouds, evaporate, thereby resolving the contradiction. "Let's give it a try," she said.

"Can't we just strike a balance between the two?" asked one of the product managers. Maggie thought he was probably worried that it might be one of his pet projects that would end up being canceled.

"We could try, but, unfortunately, compromises don't resolve contradictions. How many politicians have you heard promise to improve quality or availability of services, and then, in the next breath, claim their goal is to contain costs? Does that ever work?"

"Just the opposite," agreed most of the room.

Moving along, Maggie said, "So, what about Prerequisite 1? The trick is to ask if it's *really* necessary to accomplish our goal. Do we *really* have to increase the number of active new product projects in order to increase throughput?" While they were contemplating the question, Maggie continued. "Is our goal to have more active projects?"

One of the engineers jumped in. "Of course not, it's to complete more successful projects."

"Good point," said Maggie. "How many projects end up cancelled or don't survive more than a year or two after launch?"

"I don't know exactly," said Danielle. "But I would guess at least half."

Manny nodded agreement, before offering a halfhearted justification. "But that's the nature of innovation."

"But does it really have to be that way? And how much improvement would we see if it weren't?" There were a growing number of nods around the room.

"Does corporate care how many projects we work on?"

"Not unless they sell!" said one of the product managers.

"Manny, don't we have well more than 20 projects underway?" Maggie asked.

"That depends. Strategic level projects or tactical projects?" Strategic and tactical were DFT's terms for market-wide projects and minor line extensions or customer-requested modifications to existing products.

"Just strategic."

"In that case, we have 25 projects underway."

"And how many project managers do we have?"

"Five," answered Manny, motioning to the folks in the room.

"So each project manager is responsible for five projects?"

"About that. Some more, some less."

"How about the rest of the staff?"

"I would say they have somewhere between eight and ten projects each. But they're not working on all of those projects at once. Sometimes a project will have slow periods."

"Why?" she asked, directing her question to the project managers.

"We always end up waiting for someone else that's tied up on another project!"

"That, or we get delayed when someone gets pulled off on a tactical project or product line maintenance issue," someone else added.

"Sounds to me like a symptom of being assigned to too many projects," said Maggie, noticing nods across the room.

Suddenly, Dean, a project manager who had been quiet until this point, was on his feet, "Maggie, can I draw something on the board?"

"Certainly, Dean," she said, handing him the pen.

"As I stared at the conflict you drew, it suddenly occurred to me that something I saw in a project management course I took a few years ago evaporates your cloud." He looked at her sheepishly. "So to speak."

"Go ahead," Maggie encouraged, interested in what he had to say.

"Well, the chart I draw won't be perfect, but as engineers are assigned to more projects, the percentage of their work that adds value drops quickly. The number that stuck with me was that with five projects, you spend nearly 70% of your time on non-value adding activities.[13] Just like the list we put together earlier."

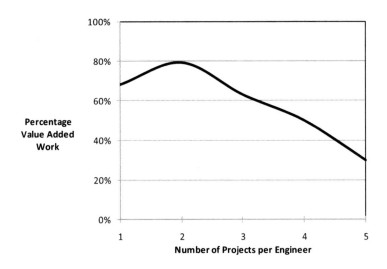

"Fascinating, Dean. What reason did the authors give for that?" Maggie was thrilled to see him so engaged.

"It had to do with multi-tasking. You know, when you get going on one project and then get interrupted for another, it takes a while to get back into that flow state again. Sometimes you even lose some of the ideas you had been pursuing." Dean paused to gather his thoughts. "So, the assumptions behind doing a greater number of projects are false. Therefore, I propose that we shouldn't increase the number of projects, but rather, we should reduce them significantly." With a flourish, he added, "I rest my case, Your Honor." The logic of his proposal earned agreement around the room. To Dean's embarrassment, a few even applauded his little speech.

"Manny, Danielle, what do you think?" asked Maggie.

They agreed, but in unison said, "You get to tell Roger and Randy."

Maggie grimaced, knowing that they had just pointed out a big obstacle to moving forward.

"You know, there's one more thing we should talk about. You mentioned that tactical projects and product line maintenance are distractions. Why is that?"

"Because we end up having to drop whatever we're doing."

"Why?" she continued.

"Usually because a customer is waiting."

"Why don't we have a small team assigned to fast-turn projects?" she asked, turning to Manny.

"Well, I guess I've always thought we didn't have enough people to dedicate any to tactical work. But this exercise has me thinking differently. Maybe a small, dedicated tactical hit squad would benefit both activities."

"Good, let's see if we can make that happen," encouraged Maggie. "Let's finish up by talking about how we are going to put all of this in action and then lunch is on me."

It was clear that the next step was to choose the projects that they would focus on. However, after fifteen minutes of debate, they still had a list of a dozen projects and were stuck on how to prioritize it any further. It wasn't something they had time to tackle today, but they were going to have to come up with a way to narrow down the list soon.

Reinforcements

Maggie returned to her office after lunch to find another message slip on her desk. *Not again*. She saw that she had missed another call from the Professor. The note said that she could reach him until 1:00 local time. Seeing that it was already a few minutes after, she hoped she wasn't too late.

She dialed quickly and held her breath until she finally heard his voice, "Hello."

"Professor, I'm glad I caught you."

"Maggie my dear, I'm sorry that I've been so difficult to reach, but I'm back in town now. I have another obligation right now, but perhaps we could meet Thursday morning for breakfast."

That would be good, thought Maggie. She was going to need some time to pull all of her questions together, anyway.

When Thursday arrived, a breakfast meeting turned out to be the perfect choice. An unusually busy morning at the Edwards' house had left Maggie with scarcely enough time for a half a cup of coffee. Somehow, she managed to arrive at her favorite little café on time, glad to see that the Professor had already ordered a carafe of coffee.

"Professor, you look well rested."

"Spending time with my two grandsons, the extreme sports duo, doesn't actually qualify as restful," he laughed while being careful not to overfill her cup. "But I enjoyed myself, nonetheless. Somehow, they talked me into hang-gliding, but that's a story for another day."

They gave the waiter their orders, and true to form, the Professor wasted no small talk. "So, tell me, how have the last few weeks progressed?"

Maggie brought him up to speed on their accomplishments, stopping when their meals were served. "So, Professor, does it sound like we're on the right track?

"Maggie, there is no track for this, at least not yet. But yes, it does sound like you are heading in the right direction. I'm especially glad to see that you took the important step of getting the team involved in making the changes they'll have to implement."

"Well, at least I remembered a few of the things you taught us at TerraGrafix," she laughed.

"More than a few, I suspect," he said. "The evaporating cloud you used to resolve the contradiction between more projects and faster projects is quite enlightening."

"One of the engineers was able to resolve it," said Maggie gladly sharing Dean's contribution.

When she finished, he asked, "So, Maggie, what is your next step to further exploit the bottleneck?"

"Somehow we have to choose which projects to work on."

"And how will you accomplish that?"

"Now, that's a darn good question. I have to tell you, though, that I'm stuck between two levels."

"How so?"

"If this were a manufacturing process, it would be straightforward. I would just schedule the bottleneck based on throughput per unit of bottleneck time required. In our case, I'm thinking that would be the return for the project per estimated hour of time in design engineering.

"And how would you measure return?"

"Well, that's a problem since few of the projects seem to meet their cost budgets...."

"That's something we should talk about, but for now if you just assumed that the budgets were accurate, how would you do it?

"Hmm...I guess I would use the projected net present value."

"There are other alternatives, but NPV keeps it simple," he agreed. "You mentioned that you were stuck between two levels."

"Well, in an oversold manufacturing environment, the orders being prioritized are for products already offered, but product development adds a new challenge because we need to decide if we should spend the time to develop them in the first place."

"And return per unit of bottleneck isn't enough?"

"No," She chuckled at the rhetorical question. "Not if you can't believe the estimate. But, I think I know how to handle it. We need to be more rigorous about assessment."

"And that's not part of the process today? I'm surprised."

"Oh, it's part of the formal process, but it's largely being ignored." She realized they were dealing with a policy constraint. In this case, though, their policy had gradually evolved into one which allowed people who were busy multi-tasking to ignore important steps.

"Of course, that would explain the large number of projects underway," he said.

"Exactly. And I'm beginning to think that our assessment method is too shallow. I don't think it goes deep enough to understand the key success assumptions."

"Why do you think that might be?"

"Hmm. My guess is it's because there are a lot of pet projects at DFT. If I want my pet project done, then I'm certainly not going to push too hard on your pet project because you might push back on mine. So, we compromise and end up funding everyone's pet projects without ever doing the kind of in-depth assessment that we should."

"Maggie, it seems to me that you have resolved your two levels. First comes assessment, then prioritization."

"I think so, but unfortunately, I'm not sure I have the tools to do the assessments."

Pausing long enough for the waiter to clear their plates and replenish their coffee supply, Maggie couldn't help but notice the gleam in the Professor's eye as he said, "I suspect you have everything you need. You just don't see it yet."

Hi-Ho Silver

Sitting alone on the top row of the bleachers, Maggie smiled as she watched Luke and Leo's soccer game. She had a perfect view of the players, as well as Jeff, who was coaching the scoreless game. It also gave her the perfect opportunity to rehash the problem she'd had assessing the 12 projects that looked like they might be a good fit with DFT's strategy.

The Professor was all set to take the projects through something he called *Guided Innovation Mapping.*™ It was a variation of the TOC future reality tree technique, specifically modified to create new product project plans. But Maggie had easily convinced him that first they needed an assessment to see if it even made sense to develop the detailed plan.

After several seconds of digging around for her notebook and pen, Maggie made a mental note to remember to clean out her purse. Finally wrapping her fingers around them, she kept one eye on the game as she jotted down some questions for the project team, realizing the multitasking was harder than she anticipated.

Jeff's voice called her mind back to the game, and she scanned the field for her boys, finding Luke at right wing and Leo at left. *"Teamwork, Falcons, Teamwork!"* Jeff's words made her laugh as she thought to herself, *That's right, Maggie; you're not the Lone Ranger up here.* She closed her notebook, realizing that taking sole responsibility for the assessment might save time, but it would sacrifice something much more important. Leave it to Jeff to remind her of the value of teamwork. It was better to wait until Monday morning when she could get Manny and Danielle's folks together. Then, they could help build the list—one in which they would feel a sense of ownership.

Creating the assessment

Week 5 – Tending to a couple of Monday morning issues, Maggie asked Nancy to notify the team and handle the details of the meeting. She had completed the most urgent tasks when Nancy buzzed to let her know it was almost ten, the earliest she could pull the team together.

The team was already seated in the conference room when Maggie walked in. Noting that Roger wasn't there, she decided to get started anyway. She knew by now what to expect. Roger would apologize later, telling her all about an important customer phone call that went longer than planned. She knew that was his way of letting everyone know how unimportant her team meetings were to him.

"Alright, let's get started so we can be out of here before lunch," Maggie said. "As most of you know, we're trying to narrow down our new product projects to a very focused list. This morning, I need your participation to brainstorm what we really need to know to prioritize our new product portfolio. So, with that said, let's start with the end in mind," she laughed at her own cliché. "Jay, what would you like to see from our new product projects?

"Well, I'm interested in making sure we get a good return. It's no fun always having finance hound us about the rate of return," he offered.

That was not the answer Maggie had hoped for. "Well, we'll get there, Jay, but the return on a project is about what's in it for us. Before there can be anything in it for us, there has to be something in it for our customers. Let's start with what's in for them. So, what questions do you think we need to answer to determine if our new product opportunities offer customers value and have a good chance of selling?"

"You're asking what our *Customer Value Lens* should be," observed Danielle.

"I guess I am," said Maggie realizing how much she liked that description. "Go ahead, Brenda. You're next."

"We need to make sure that we're not solving a problem that someone else has already solved." That was an excellent observation. Glad that she'd made this a team effort, Maggie asked for clarification. "Do you mean like a 'me too' kind of product?" she asked.

"Yes, but it's more than that. Sometimes, it's almost like we fall in love with our technology as a way to solve problems. But if there is already a good, cost-effective solution, we're going to have a hard time selling the new one."

"So, you're saying that we need to make sure our new products solve an unmet need," said Maggie.[14] She was pleased to see they were off to a good start.

"Yeah. I think that sums it up."

"Alright, let's keep moving," said Maggie, while quickly writing Brenda's comment on the flip chart. "Danielle, you're next."

"We need to know what job the customer is hiring the product to do," she answered.

Realizing this was going to slow them down, Maggie wondered if she was getting more than she'd bargained for. She immediately recognized the key concept from the *Innovator's Solution*,[15] a book that Danielle had shared with her. But now that Danielle had brought it up, she needed to make sure the rest of the group understood the point. "Okay, we're not going to debate the merits of each of these, but for the sake of clarification, could you explain what you mean by jobs, Danielle?"

"Sure, '*Jobs to be done*' is a market segmentation framework that Clayton Christensen developed to help find new opportunities and market segments. For instance, are customers buying water filtration? Or are they hiring us for reassurance that the water supply is going to be clean and safe—or maybe to reduce their energy usage in filtration?"

"What's the difference?" asked Dean.

"The different types of jobs that buyers have create significantly different needs. Any small micron sieve can provide the filtration, but a multiple redundant filter element with a system for on-line monitoring gives them assurance—peace of mind."

"So, there might be a segment that buys on security of supply."

"That's right, and their price point is likely to be different, too."

"Okay, we've got that one." Maggie looked to the next person.

"How about knowing that our solution creates value for the customer? And after learning a little bit more about TOC, I want to know what value it creates for the customer in terms of T, I, & OE.[16] More Throughput, less Investment, or less Operating Expense."

"Finance will like that one," joked Jay. "But they're also going to want to know whether the value is large enough for us to make some money, too."

They'd gone around the room once, and it was Danielle's turn again. "I'd also like to know if it is a sustaining improvement or a market disruption."

"Whoa there, now, you're going to have to explain that one, too. I hear that word disruptive anytime anyone wants to sell a new idea," said Manny. "Frankly, it gets a little tiring."

"Doesn't it just mean a breakthrough?" asked Dean.

"Well, it's almost universally misused," Danielle agreed. "Basically, a disruptive technology means that you've found a new basis for competition in an over-served market. I know that sounds a little too much like consultant speak, so let me give you an example. I'm sure you've all seen one of those commercials where they use a vacuum cleaner to pick up a bowling ball."

"Sure."

"Well, how do most new vacuum cleaners compete?"

"Performance, I guess. Higher suction, bigger motors, even high temperatures to kill bacteria."

"In general, though, what's happened to prices even with all that performance and all those features?" she asked.

"You can get a lot more for your money. That's for sure."

"That's a sure sign of an over-served market. More and more features and value are added without a higher price."

"I see us doing that sometimes, too," said Jean.

"Well, it was a way of life in the vacuum market until the Roomba® came out. I know you've all seen these little robotic vacuums in that internet video that was going around the office—the one where the dog keeps attacking the vacuum." That elicited a few chuckles. "Well, iRobot® used the jobs to be done approach we talked about earlier and found a new basis for competition. It's no big surprise, but no one likes to vacuum. The competition was offering more vacuuming power when what customers really wanted was freedom—freedom from the task of vacuuming. Roomba created an entire new category based on the convenience of unattended cleaning. That said, did they compete on the traditional basis of performance?"

"No way; I've got one of them, and it will never pick up a bowling ball," piped in one of the project managers.

"But you still like it. You didn't return it because it lacked performance."

"Are you kidding me? The thing picks up everything I can see, and I never have to bother vacuuming. What's not to like?

"So, the performance is good enough for the job to be done," concluded Danielle.

"Absolutely."

"So, what's the connection to our assessment, Danielle?" interjected Maggie.

"There's a very high failure rate when companies do one of two things: number one, when they try to develop and market a product that's disruptive to their own market or number two, when they try to enter a market leader's space using a sustaining strategy. Because of those high failure rates, we should avoid projects that try to do either.

"You're going to have to explain that again," said Manny.

"Sure, years ago when they created the PC, why do you think IBM created a physically separate division which reported only to the CEO?"

"Because the mainframe guys would have blocked them at every step."

"That's right. Wasn't there a guy at Digital who was infamous for saying no one would ever need a computer at home? Oops," someone else added.

"Exactly, organizations are designed to run efficiently by serving today's customers, and as a result, they starve disruptions," said Danielle.

"So, it would have been a mistake for Hoover to try to bring out a robotic vacuum?"

"It's very likely that they would have struggled because the organization was built to do something else. They probably would have obsessed over performance and then gotten the price point or the usability wrong—not because they're no good at product development, but because they're hardwired to see power as the performance driver, the same way the mainframe guys were."

Danielle moved to the remaining issue. "Now, as far as trying to enter a market leader's space with a sustaining improvement, does anyone know what happened when all of the internet only banks came out and were going to put traditional banks out of business?"

"The brick and mortar guys all invested in web technology and started offering electronic banking."

"That's right. They defended their position. The only way to attack someone else's market is with a true disruption. It has to be a new dimension that they either can't or won't compete on."

"I can see how that fits," said Maggie.

After several more trips around the room, Maggie sensed they had exhausted the topic, so she thanked everyone and sent them on their way. Impressed with Danielle's energy on this topic, Maggie asked her to pull together a draft framework. She could tell this was going to be invaluable in making sure that only the best projects made it into their bottleneck.

Rough start

Week 6 – Monday morning came earlier than usual, since Maggie had to catch a 6:00 flight to Nebraska. She was disappointed that it had taken her this long to get a meeting with one of DFT's largest industrial customers, Midland Enterprises. But, at least they had agreed to meet with her and Roger. While there, she received more than enough information to confirm what she already suspected. Midland had been looking at competitive alternatives because DFT had been so late with their new systems. They were going to need to get the upgraded Energy-Saver system out by year-end if they were going to keep their business. While it wasn't what she wanted to hear, Maggie thanked them for the feedback. While she couldn't promise she could fix the issue, she assured them she was trying and explained some of the changes they were making to get this important program back on track. She was glad that Roger had been there because, regardless of his other faults, he was gifted when it came to soothing customers ruffled feathers. But at the same time, she knew that Randy would get an earful, too.

Unfortunately, while she was out of town, a crisis erupted back at the plant. A primary supplier had a major fire in its facility, which was likely to cause a supply interruption for several weeks, if not months. So, when Maggie returned, she spent several days with Jim Hollister, her VP of Manufacturing, evaluating the situation and confirming that the contingency plans they already had in place would suffice. Jim's team had done a solid job of planning, but it was still too early to tell.

Top five

When Friday arrived, Maggie realized what a blur the week had been. She was pleasantly surprised when she checked her email late that afternoon to find that Danielle had not only pulled together the draft assessment document she had requested, but had also worked with her folks and the project managers to fill one out for each of the top 12 projects. They could easily set priorities with the information they had.

Even with the progress that Danielle had made on the assessment, she still felt frustrated, given the events of the week. One of her major customers had given her the gift of clear and honest feedback. But, she wasn't sure whether she could move the organization quickly enough to change the situation. It would be Monday before she and Manny would have time to sit down and discuss any steps they could take to try to complete the project sooner. She wondered what every day of delay in a new product launch was costing them—but knew it was probably more than anyone realized.

Readers can download a copy of the project assessment worksheet at www.SimplifyingInnovation.com/extras

Risk reduction

Week 7 – Maggie was finishing a report for corporate when Manny stuck his head into her office late Tuesday afternoon, "Can I interrupt you for a minute?" Seeing that she was busy, he said, "Oops, stupid question. I see I already have."

"No, no, come on in. I need to clear my head, anyway. What's up, Manny?

"Well, I ran across some very interesting data that further confirms the value of doing good assessments."

"Let's see what you have."

Manny handed her an article, her eyes drawn to a graph brightly circled in red.[17,18]

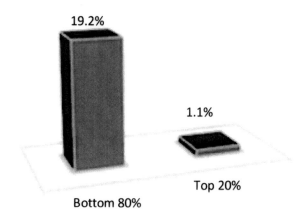

Percentage of Projects Cancelled
after Detailed Design

"This is from a different industry, but the graph shows two different groups. The top 20% of performers, on the right, cancelled very few projects once they began detailed design—just a little over 1%. But the bottom 80% cancelled nearly 20% of projects after spending scarce detail design resources. Now, the top performers cancelled just as many projects. It's just that they cancelled them in the assessment or feasibility steps before they ever got to the design stage."

"Wow, that's pretty dramatic!" agreed Maggie.

"That's not even the best part. The top 20% achieved double the profit from new products and managed to complete projects in half the time that their competition did," he said. "And remember my concern that we could become too risk averse?

"Sure."

"Well, I've realized that's part of what's so important here. High performing innovators take just as many chances initially. It's just that they use assessment to weed out the unnecessary risks and then concentrate that extra effort on the promising ones.

"Unnecessary risks?"

"Yes, why take unnecessary risks if assessment or feasibility work could tell you in advance that they have a high likelihood of failure?

"Manny, this is great stuff. Make sure you share it with the rest of the team."

"Definitely. If there is any doubt left about the need to run fewer projects and conduct more rigorous assessments to fully exploit our constraint, this should help erase it."

Strange brew

"Okay then, Maggie. Tell me about your progress," the Professor said as he poured Maggie coffee at the café that had become their usual Thursday morning meeting spot.

"Well, Professor, we've been working on the Exploit step, and we now have assessments to filter out bad projects, prioritization to choose between the good ones, and metrics to identify how well our process is working. TOC is helping address the process element. How can we manage both process and project?" That was a complication she wasn't sure she was prepared to deal with.

"What contradiction does that create?" he asked.

"I'm not sure it's a contradiction exactly, but it seems to me that while we need to continually try to get more out of our process by attacking the constraints on the bottleneck, we also need to continually improve in our efforts to get each project done in as little time as possible."

"Maggie, you've hit it spot on. The way in which you plan and manage projects can also be one of the constraints impacting the bottleneck."

"I guess that makes sense."

"Maggie, remember when we first spoke? I asked you what the goal of new product innovation should be, but also how you would measure progress against it. Isn't that what we're talking about now?"

"So, if the goal of new products is to make more money in the future, which we measure by new product throughput," she pondered, "and we need to improve both our process and the management of projects to do so...Then, I guess the metric for that would be new product cycle time."

"Yes, but you must be explicit. Throughput is a defined term, but what is your definition for cycle time?"

"Well, I would say cycle time is the length of time it takes for a project to go from approved proposal to generating cash flow?"

"There you have it, then. You could even estimate break-even—the time required for sales throughput to offset the investment,[19] but I think you're better off to start simple, like you have. I'd suggest you write that down and make sure you have agreement internally."

"I'll do that."

"Maggie, do you realize how powerful that is? You now have two simple metrics, new product throughput and cycle time, that can tell you whether both your innovation process and the projects you are running within it are improving or degrading."

"I guess maybe that's the problem I'm struggling with. I know how TOC can help us tackle the process improvement, but I need a framework for the project part."

"Fortunately, Maggie that's already been done for us," he laughed. "Remember when I tried to tell you about Critical Chain and our phone call was cut short?"[20]

"I didn't realize it was Critical Chain that you were talking about. Isn't that another of Dr. Goldratt's books?"

"Indeed."

"So, isn't that just the critical path that's taught in most project planning?" asked Maggie.

"Heavens, no. It's much more than that. The critical chain is the project's constraint, or the longest sequence of tasks comprising the project's cycle time or completion date, taking into account constrained resources. Traditional critical path methods only work in the unrealistic case where there are no resource conflicts. We know that rarely happens, which is why most projects finish late. Otherwise, we wouldn't need to have this conversation."

"That's for sure," she laughed.

"Critical Chain Project Management addresses the conflicts ignored in traditional planning approaches. It's grown into a very robust set of tools. Interestingly enough, Maggie, you've uncovered some elements of it yourself."

"We have?" She was surprised.

"Yes, the evaporating cloud you developed about interruptions is a big part of that. In fact, the general concept of eliminating bad multitasking is part of Critical Chain."

"Wasn't there a section on this in the resource manual you put together for us at TerraGrafix?"

"Yes, but we never got that far in your TOC implementation."

"Now I'm definitely going to have to take a look at it."

"Maggie, would you be willing to let me come in and work with your team to implement this?"

"I thought you'd never ask, but as you know, cash flow is an issue for Barrister these days."

"Maggie, you've been a great friend and client over the years, but I'm doing this because I'm intrigued. Besides, I suspect that what we are developing here is going to have far wider applications."

"So, how can we get started?" she asked, realizing this was a critical step in getting the Energy-Saver project on track.

"Actually, I was thinking that we should include an overview of the concepts in one of the first two sessions I'd like to do with your team."

"The first?"

"Yes. Since your team has had so many distractions, what policy constraint could be affecting the projects plans that are in place now?

"Well, I would imagine that we've allowed teams to work with plans that aren't very thorough or might not even be feasible. And that's for the cases where we actually have plans," she admitted.

"Exactly, similar to what you mentioned with the assessments. So, we'll keep it simple. One training session needs to focus on how to identify the right tasks, and the other needs to give them the tools they will need to create a robust plan. I have to warn you, though, that after the large group session, we'll still need additional sessions with each project team to identify the particular obstacles for their plans."

"Obstacles?" Maggie was wondering how they would fit it all in, but she also knew that every hour of planning could save many, many hours in execution, so she didn't object.

Can't say I didn't ask

Maggie was frustrated. She had been chasing Roger for nearly two weeks, attempting to get his input on project priorities. She could understand that he wasn't happy about being sidelined, but he was becoming an obstruction. It was Friday morning before she finally cornered him in his office. "Roger, I'm going to need your input on new product priorities, or we'll have to move without you." She also wasn't pleased that she had to spend 20 minutes explaining the changes that the team wanted to make. Roger had been invited to the sessions, but he always seemed to find some last minute crisis as an excuse. Customers were important, but she knew there was more behind Roger's absences than that.

"Maggie, we need everything that R&D is working on. These are all things customers are telling us they need."

"Yes, but which are most important?"

"They all are," he insisted.

Maggie thought that maybe she needed to work with Roger to help him understand why having the engineers working on too many projects was an issue.

"We tried this participative crap before and it didn't work that time either." He said, trying one of the oldest blocking maneuvers around.

She tried to walk him through the goal discussion that had been successful with other team members, but he quickly put up a defense. "Of course, the goal is to make more money, but I sure don't see how giving the folks in R&D a vacation is going to help with that." She tried to engage him to discuss the twelve projects, but it was clear he didn't want to see any changes.

"Well, here's the list that marketing generated as being the top twelve projects. We're trying to pick five of those to work on at one time."

"Five? We have over 60 projects running! What you are proposing would put us out of business!"

"Roger, I'm just talking about the strategic projects. We're going to carve out 25% of the engineering group to work solely on product support, tactical line extensions, and customizations. The other 75% will focus on the strategic development projects. Besides, Roger, I recall only a few weeks ago you said that nothing was coming out of R&D. Wouldn't you rather have five new products completed, instead of 25 projects moving along at a glacial pace? Plus, based on our history over the last few years, less than half of those projects will be successful. So which do you see as most important and why?"

"Maggie, you do what you think you have to, but take it from me—you're heading down the wrong path. I just don't want to see you make a mistake like this."

Maggie finally threw her hands up in the air and said, "Alright, Roger, we're going to move forward without your input, but you can't say that I didn't ask."

Keeping Doug in the loop

Later that afternoon, Maggie and Doug met with Jim Hollister and Dynamic's purchasing manager. The supplier outage had been resolved with no downtime, but the team had still learned a few things from the experience. As a result, they were proposing some changes.

After they had finished and left, Maggie decided that she needed to talk to Doug about her struggle to get Roger's participation in the changes she was making. She hated to talk about others when they weren't present, but she also didn't want any surprises. She was very careful to keep the discussion factual and professional. Doug listened, then simply sighed and encouraged her to continue on the path she had started. "We both know it will lead you to the improvements we're looking for." Clearly, this was part of the reason that Doug had made the changes he had. But Doug had a boss, too, and although she knew he'd never raise the issue in front of Maggie, Roger's connection to Randy had to make the situation tenuous for him, as well.

Finding obstacles

Week 8 - When the DFT team arrived at the conference center for the first session with the Professor, they saw that Maggie's rules about electronic distractions were still in effect. But this time, no one complained. Actually, the last meeting had been one of the most productive they had experienced in some time, and no one wanted to argue with success. For the first time in years, they knew what their top handful of new product priorities were and were able to focus on them.

With everyone seated, Maggie went through the morning's agenda and introduced the Professor. "I'm here to learn, just like the rest of you. So, I'm going to turn it over to our good friend, the Professor, and then try to stay out of the way."

"In your last session with Maggie, does anyone remember someone raising the issue that improving new products involves both projects and process?" Receiving the acknowledgement he was looking for, he continued. "Well, I know that Maggie has been working with you on implementing the five focusing steps of TOC to improve your innovation process. But today as part of the exploit step, we are going to focus on the project side. We're doing that because you identified two of the issues constraining your bottleneck as a lack of in-depth project planning and understanding what constitutes feasibility. Now, Maggie referred to today as a training session. I have to admit that it isn't going to be a training session as much as it will be a doing session. We're going to train by developing the Guided Innovation Map for one of your projects."

"Manny, which project are we going to use as an example?"

"The ultra compact emergency filtration system," answered Manny. "Everyone should have a copy of the project assessment in front of them, and I hope you've all had time to look it over."

"Very good," said the Professor. "Then can anyone tell me what the goal of this project is?"

"To launch a new product with an ultra compact footprint design for emergency use by government agencies."

"Why is that important to your customers?"

Several people in the group started to throw out features and benefits.

"Please, let's focus on the one big benefit—the problem you are solving for them," he redirected them.

"It would have to be to provide a water filtration unit that can be easily transported into remote disaster areas without expensive airlifting—a unit that FEMA or the National Guard can bring in using just an off-road vehicle," Manny answered, before asking, "Out of curiosity, why list just one benefit when there are others?"

"I'd wager that if your product didn't deliver that big benefit—solve that big problem, it wouldn't be very successful. Would it?

"No, I guess not." Manny agreed.

"You see, it's important to make that big benefit part of your goal so you won't ever lose sight of it. That's important not only during development, but later during marketing, as well. Now let me ask, why do you want to solve this problem for them?"

"To make more money, of course." Maggie smiled at the answer. It showed the team had been paying attention, but she also knew the professor wanted more.

"Let's hope so," the Professor kidded, "but please be more specific."

"We think we can reach about $30 million in system sales by year 5."

"Now, can you turn all of what we've covered into an objective statement?" He gave them a few minutes, and the group finally arrived at:

"Launch an ultra-compact emergency filtration system that can eliminate airlifting and will generate $30 million in system sales ($5M in cash flow) by Year 5."

"That's much better as long as you have an agreed definition for ultra compact. That is your high-level project objective." He drew a bubble at the top of the flip chart and wrote the objective in it. "Now, what has to happen in Year One, if you are going to reach that high-level objective?"

"We have to launch the product and start selling it."

"Okay." He drew a second bubble below the first and wrote 'Year 1 launch' in it. "Now I'm going to ask you an unusual question." He paused to see if he had their full attention. "What are all the possible reasons you can dream of that could somehow cause this project to fail?"

A look of discomfort fell across the team. "You want us to talk about failing?"

"Let me explain. We are going to spend some time brainstorming every possible obstacle to the success of this project. Then, we'll assess which are true obstacles that we must address. Finally, we'll develop step-by-step solutions for getting around those obstacles and that will serve as the starting point for your project plan. When we're done, you'll have a map which will give you a full listing of all of the tasks that must be completed and in which order." He paused while they considered that and then asked the question again. "What are all the possible reasons that could somehow cause this project to fail? Jackie, would you mind starting?" He thought it best to get the project managers engaged early.

"Hmmm...I want to know what the flow requirements are."

"But what is the obstacle?" the Professor asked.

"That we don't have the necessary understanding of the flow requirements."

"Excellent, Jackie. I'll write that down. Who's next?"

After five rounds, the group had exhausted the obstacles. After a short break, the Professor moved them on to the next step. "Now, the first question I have is whether each of these is really an obstacle. Let's start with this one from Jackie: 'We don't have the necessary understanding of the flow requirements.' Is this really an obstacle? Is it possible to design the system without it? Please, somebody other than Jackie."

"Absolutely not."

"Is the information available elsewhere?"

"No, we'll have to do some testing to define the requirements."

"Okay, then it is an obstacle. Does everyone agree?" With the group's consent, he continued. "The next step then is to turn it into a task. Anyone?"

"Wouldn't it just be to complete the testing to define flow requirements?"

"Exactly. And that's how we handle each obstacle."

"Jackie, please write that task on one of these colored sticky notes. Then, hang onto it for later. But first, is it a technical, a commercial, or a manufacturing task?"

"Technical."

"Then, use an orange one for technical. Commercial will be green and Manufacturing will be blue."

"Let's try another," he said, choosing, 'We don't know how the new design will function in the field. '

The group agreed that was a real obstacle and something they run into often.

"What task would get you around the obstacle?"

"Beta testing a number of units in the field before commercial launch. It's a common industry practice, but it's something we often rush or even skip when projects are late." said Danielle. "Usually with cost consequences when we have to take units back or upgrade them in the field."

The Professor asked Manny to lead the group through the rest of the list while he observed. They continued through the list until they came upon 'Not enough testing or prototyping resources.' "Is that really an obstacle?" asked Manny, approaching it as he had with each item on the list.

"Sure, they're always busy and never seem to get to my projects," said Brenda, another project manager.

"Do they not have enough capacity or was it a matter of priorities?" asked Manny.

"Not having enough resources or enough funding commonly comes up as an obstacle," said Manny. "But is it really an obstacle?"

"It is if you're the project manager." This response from Dean elicited Brenda's agreement.

"Dean, isn't this a concern based on the way we were operating before with too many projects competing for resources?" asked Maggie. "We need to build the plan, identify the resources needed, and manage the plan from there. If the resources are not available, then we should revise the plan to reflect that or delay implementation until they are available."

"Hmmm... I guess I can live with that," said Dean.

Up to this point, Manny had moved them along satisfactorily, but the next obstacle proved to be quite a hurdle.

Hinge assumptions

"The next obstacle is *'We haven't proven that we can increase the payload enough by operating in a subsonic flow regime'*," said Manny "Is this really an obstacle?" he asked with a sheepish look and tone. It was clear that he was still trying to get comfortable in the role of facilitator.

"Absolutely," was the resounding answer. And the task was *'Prove that operating in a subsonic flow regime will increase the payload by at least 50%.'*

"Professor," asked Manny, "what happens if you can't turn the obstacle into a task?"

"It is always possible to create a task, but the task may not be possible to accomplish."

"I guess maybe that's what concerns me," said Manny. "This particular task could be a tough one. We have some interesting lab data, and there is theory to support it, but we've never seen that level of increase."

"In fact, that is why I call this approach *'Guided Innovation Mapping'*. Using this process will help you identify the high-risk tasks, and it will guide you to the places in the project where there is some physical contradiction that we must innovate or invent around."

"Could you give us an example of what you mean by innovating around a contradiction?" asked one of the project managers, whose look of bewilderment made Maggie wonder if he was alone or if the others had the same level of confusion.

"Yes, it's what happens when we need more strength out of a structure, but we also need it to weigh less. Strength and lightweight are contradictory properties so we need an invention. Does anybody know of one?"

"How about aluminum or magnesium?"

"Yes, but let's assume that we've already done that and still need our structure to be stronger or lighter, or both. What else can we do?

"We used carbon fiber composites when I worked in the defense industry." said Brenda, "I could tell you more, but I'd have to kill you all afterwards," she said with a sly grin.

"No, that's quite enough information, I'm sure," laughed the Professor, "and a good example of an invention that solves contradictions."

"But we wouldn't know if the carbon fiber was enough without doing some more design work and extensive testing."

"I'm glad that you raised that point, Manny," said the professor, "because we need to make sure that the project assessment tool includes explicit statements of all of the hinge assumptions."

"Hinge assumptions?"

"Yes, the limited number of critical assumptions upon which project success hinges."

"Are you just talking about technical assumptions?"

"No, they do include technical assumptions, but also include critical commercial and manufacturing assumptions that define feasibility. And depending on the industry, regulatory and intellectual property can sometimes be hinge assumptions, too."

"Well, it sure sounds to me like this project hinges on being able to achieve the 50% flow increase," said Jackie.

"And the biggest hinge assumption on my Energy-Saver project is going to be that the cascading flow design will give us the savings we expect," added Dean.

The project map

Having converted all of the obstacles into tasks, it was time to build the map. The Professor went back to the original flip chart with the project objective and then drew the following framework on the large whiteboard:

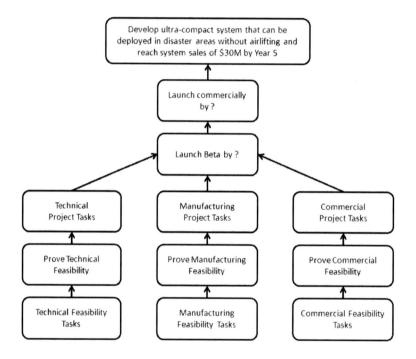

"So, the next step is to take each of the tasks and place them in the general areas where they belong and then determine which need to be done in what order. In another session we'll finish up by assigning task estimates and turning each of these maps into final plans."

"Professor, the structure of the map seems to indicate that there is work going on before feasibility is proven. Doesn't that mean that bottleneck resources could be wasted?" asked Manny. "I thought that's why we did the assessment work—to make sure that didn't happen."

"Indeed," said the Professor. "So, how would you manage that?"

"I'm thinking," said Manny studying the map intently, "that it would make sense to try and complete as many of the non-bottleneck tasks as possible first."

"Why?"

"Well, because if any of the non-bottleneck tasks disproved feasibility, we'd stop the project and move resources to another. That would minimize use of bottleneck resources on unproven projects."

"But, Manny, that seems to me like we're shying away from taking risks. Winners in innovation aren't afraid to take risks," countered Dean.

"I agree, Dean," said Manny. "I struggled with that same issue until I understood that winners figure out the risks worth taking. Why take risks that you know will fail when some upfront feasibility work can tell you whether they are worth it?"

"Okay, that makes sense," said Dean. "If it isn't feasible, there's no reason to take any further risk. I can buy that."

Maggie looked at the Professor and winked, pleased with the leadership Manny was taking.

Bright shiny objects

Week 9 – That Monday, Maggie invited Danielle to lunch as a way of saying thanks for pulling together the project assessment. While they sat in her favorite Thai restaurant and shared an order of fresh spring rolls, Maggie said, "Danielle, I really want to thank you for the work you did to put this together. I can see how passionate you are about the market-focused elements of innovation."

"Well, thanks Maggie, I really appreciate that coming from you. It's been such a great learning experience seeing this all unfold. You know, I've been spending some time working on a plan to extend the approach we developed to drive improvements in actually finding more new opportunities."

Maggie asked her to explain what she was hoping to accomplish. Danielle described how she wanted to institute a program she called Customer Value Lens™ that had several elements, including additional training on customer visits and understanding customer value. Her objective was to find more and better opportunities for DFT. She even drew a little diagram showing how different projects would be evaluated.

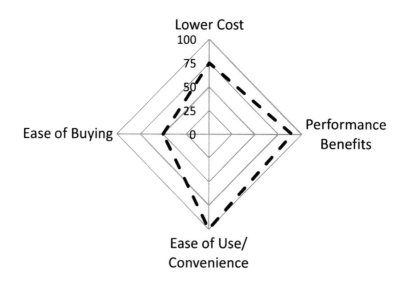

"Maggie, let me give you an example using the new high uptime system planned for launch in Q2." She turned over one of the extra placemats on the table and drew a small chart. "The upper axis is Lower Cost, while the lower axis is Ease of Use/Convenience. The left axis is Ease of Buying while the right axis is Performance Benefits.[21]

Fifty is average or current performance. So if we look at our system, it scores pretty well on reduced cost. While it actually costs more than other filtration systems, the lower installation and lifetime maintenance costs give it a strong advantage. As far as ease of use, it scores well there, too. It's fully automated, even the filter regeneration cycles. It also has error reporting that can actually send an email or text message to the operator so it doesn't require their attention. From a buying standpoint, there's no real benefit because our competitors also sell pre-engineered systems. Finally, on performance, there is a big advantage in uptime and total output. Because of the dual redundant system, the rated filtration capacity will always be available with no downtime."

Maggie stared at the diagram on the back of the placemat and said, "Wow, Danielle that really helps me see what you're talking about. We should definitely include this as a part of the project assessment." Maggie picked up the diagram to make room for their Pad Thai, which had just arrived. "Can I keep this?"

"Sure, and we should definitely include it in the assessment. But what do you think about rolling all of these concepts out more broadly so we can use them to generate more new product ideas?"

"Well, Danielle, I guess my first question would be what would it take to implement something like that?"

"That's a fair question. Well, we'd have training for the new product teams and the marketing people. Quite honestly, to be able to use these ideas, we'd also need to train the teams on how to interview customers and then how to recognize the unmet needs that represent the opportunities."

"That's what I would have guessed, too. Unfortunately, I don't think now is the time."

"Maggie, I'm confused. You said..."

"I said that I think we should make sure the assessment includes all the elements as part of the critical quality check on any project we release to the bottleneck. But let me ask you, where is our constraint?"

"In Design Engineering."

"That's right, and what will implementing this idea do to get more through Design Engineering?"

Danielle chewed at the edge of her lip, straining to come up with another argument for the idea, "I'm struggling here, Maggie. If it's a good business practice, why shouldn't we implement it?"

"Danielle, it's a good practice, but it also requires an investment of time and resources to implement. And what will the return be?"

Danielle laughed. "Alright. I get it. With everything I've read on TOC, I still overlooked one of the important assessment elements we can put in place as managers."

"What's that?"

"If it doesn't increase the system's throughput, don't do it."

"That's generally right." Maggie was pleased with the progress Danielle was making. "Using these concepts in assessment is critical to fully exploiting our development bottleneck, by making sure only the best projects get resources. But at this point, we don't necessarily need to improve on idea generation because we have more ideas coming in than we can currently handle.

Eventually, we'll want to take this step because it will help protect the bottleneck by ensuring a flow of opportunities into the pipeline—so-called protective capacity. But, for now, I'd like our focus to stay on successful execution of the product launches we're facing. Your team needs to be fully ready to move forward when Engineering delivers the products."

"Rookie mistake, I guess."

"Not at all. It's very easy to become inflicted with what the Professor calls 'Bright Shiny Object Disease,' regardless of your experience level."

"I beg your pardon," Danielle laughed.

"I know, it sounds terrible doesn't it? It simply means that, as managers, it's quite easy to become distracted by all of the good new management ideas or even technologies that come along."

"We used to call it 'Program of the Month Syndrome' at my previous company," she admitted.

"That's right, but most of those ideas went by the wayside, didn't they? Some people would say that it's because management didn't wait long enough to see results," Maggie reflected. "In some cases, that's true, but in 80%, I bet they weren't addressing constraints, so the results were never going to be there, anyway."

"That's right, the *Pareto Principle* tells us that the top 20% of programs will generate 80% of the results and the bottom 40% will generate almost no results.[22] But, Maggie, I think the effect with TOC is even more pronounced than that. Most of the ideas focused on improving non-bottlenecks won't generate results at all."

"Wow, I never thought of it quite like that before, but since new approaches are bound to come along, TOC provides a framework for determining whether they deserve our attention—helping determine if that's where we need to put our improvement efforts. If it doesn't increase or protect the system's throughput, don't do it."

At least they had themselves convinced.

Playing defense

A few days later, Randy Barrister surprised Maggie by popping his head into her office. "Hello, Maggie. I was out this way for a lunch meeting and thought I'd see how things are going."

Walking around the desk to shake his hand, Maggie said, "Hello, Randy. If I'd known you were coming..."

"No, that's okay, Maggie. Nothing formal. I just wanted to hear what kind of progress you're making," he said, pushing the door closed.

"Sure, no problem. Have a seat," she said while she pulled out a chair at her conference table.

Never one for chitchat, Randy jumped right in, "Of course, I am a little concerned by some of what I'm hearing about cutting the R&D project portfolio, but not reducing your staff."

Now, who could have been complaining to Randy when we've hardly even gotten started? Maggie realized Roger must have been whispering in his ear already.

"Randy, I am focusing the team's efforts on a much smaller number of projects, but only so that..."

"Maggie, I'm not sure that I support doing fewer projects. I'm told we need all of them to deliver what customers are asking for," he said, confirming Maggie's suspicion about Roger being the impetus for this visit. "Look, I'm just going to come out and say it. When Doug wanted to put you in this position, I told him that it was a risky move—maybe even a mistake. You did a great job at TerraGrafix, but this isn't manufacturing." She knew he wasn't a fan, but his words still stung.

"Randy, when I took this job, I had some of those same concerns. What did I know about new products?" It was a risky thing to admit, but she always believed in being direct. "But what I do know is continuous improvement of processes. And that's why I accepted Doug's challenge. It's also why I believe we can make real improvements." She paused briefly to let it soak in. "At TerraGrafix, we were able to increase throughput by cutting both lead times and inventories," she continued. "If you remember, that was considered risky, too. Sales and customer service were sure that it wouldn't work. But we would have never taken a risk like that unless we knew it was minimal. The common sense of TOC and a cause and effect approach showed us that it was the only option."

"Hmmm...what does inventory have to do with new products?" Maggie got the impression he was sizing her up.

"A lot actually," she said. "Having too many projects in our pipeline was preventing us from fully exploiting our constrained resources. It was actually causing the lead-time for all of the projects to increase—just like increasing WIP inventories increases manufacturing lead times." Maggie went on to explain some of the logic behind fewer projects and decided it might be helpful to share an example. "Randy, let me show you why doing fewer projects at the same time, or pipelining projects as it's called, actually increases sales throughput." She went up to the board and drew a graph of cash flow vs. time.

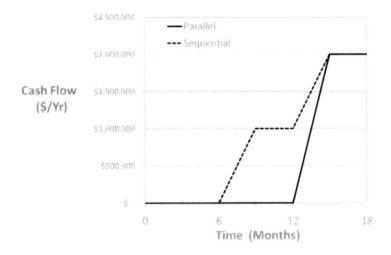

"If we have ten people available to work on two projects, both of which will take a year for five people to accomplish and will generate $1 million per year in cash flow after launch, what should we do?"

"Put five people on each, but what does this have to do with..."

"I'll get there, I promise. So, both projects finish at the end of 12 months and ramp up to deliver a little less than $2 million in total cash flow over the second year. Assuming sales ramp up in 3 months from launch, just for the sake of this example, the cash flow for your parallel approach looks like this."

"Now, look what happens if we take a sequential or pipelined approach instead and put all 10 people on one project at a time."[23] She continued, "Project 1 finishes up six months earlier and begins delivering cash six months earlier."

"Maggie, you can't necessarily double up the number of people—they'd be tripping over each other."

"Point taken, but Randy, the projects we are doing aren't operated at anywhere near full staffing. With 25 projects, the average engineer is working on seven projects. It's more likely that by doubling up on resources, we'd see projects done in even less than half the time because we'd be eliminating all of the inefficiency of waiting time."

"Go on," he said, indicating some interest for the first time. The number of projects per engineer seemed to be new information to him.

"Project 2 follows right along and, at the end of the first year, begins delivering its promised cash flow. And the important part here is the area between the two curves which is the extra cash flow. Focusing on one at a time, doing them sequentially, generates an extra half million dollars of throughput and both projects are still done at the end of the year. The customers for Project 2 are no worse off and the customers for Project 1 are happier."

"That's surprising," he noted, quietly contemplating the example. "If that worked, it would also reduce risk."

She hadn't thought of that, but he was right. Doing any given number of projects sequentially rather than in parallel reduced both the time to payback and risk. In this example, it added an extra six months of strategic flexibility before having to commit to Project 2. Had she won him over that easily?

"But I need more than theory, Maggie. I need you to deliver. Innovation is important, but my priority has to be cash flow. If I don't start seeing new products heading in the right direction, we're going to need to talk about cutting costs."

Maggie knew that the cost cutting would include her if she wasn't able to get things moving, and quickly.

Critical chain

Week 10 – The DFT team was once again seated around the conference center table. After completing the Guided Innovation Mapping for each project, they were excited to get started, anxious to see how TOC could help with managing projects. Since they already knew the Professor, he kicked off the session. "Last time, we talked about the fact that driving new product improvement involves both process and project improvement." Noticing several head nod in agreement, he continued, "Well, today we're going to talk about using Theory of Constraints to improve project management, something that's also known as Critical Chain Project Management." Getting right to business, the Professor asked, "So, how many of our projects finish on time?"

A wave of groans swept across the room. It was clear that Murphy had struck all of them at one time or another. "Don't we build in buffer time to compensate for this?" he asked.

"Sure, but you can't predict what problems you might run into." Manny's explanation was accompanied by the agreement of the project managers, who were fervently nodding their heads.

"I see. Then, let me ask how many of you see projects finish early?" No hands went up. "If it were simply an issue of variation, wouldn't some projects finish early?"

"I'm all ears," said Manny, ready to hear more.

"Indeed, variation and resource conflicts are definitely problems, but there's another problem that magnifies their effect. You've all studied at a university at some time in your life. Let me ask you a question. If we give most college students two weeks to complete an assignment that can be done in one evening, when will most complete it?"

"The night before," was the resounding reply.

"Of course," he said. "And that 'Student Syndrome', my dear friends, is almost certainly part of our problem."

"I don't understand," Maggie interjected. "Isn't the slack or buffer that you mentioned built into the project to allow for that?"

"No." This time, Dean was leading the charge. "The buffer is supposed to allow for variation. You know Murphy's rule: Whatever can go wrong will go wrong. But since it is used up on procrastination, any real variations, such as resource availability, end up truly delaying the project. When you understand that, it really explains a lot."

"Okay, Professor," said Maggie, joking with Dean. "I get it now."

"Very well put, Dean," commended the Professor. "But now, what do we do about it? Any ideas?"

"Watch your project team members like a hawk?" offered one of the product managers.

"Yes, some percentage will toe the line if you stay on them. But how effective do you really think the hawk approach is and for how long can it be sustained and still be effective?" asked the Professor.

"Only as long as you're watching," the product manager agreed.

Looking across the group, the Professor threw out another scenario. "What if we don't give the tasks any buffer in the first place?" Then, he crossed his arms and stood back, waiting for the inevitable reply.

"You can't do that. What about variation? Murphy always sneaks into every project."

"I agree, I agree," said the diminutive professor, raising his hands to fend off the onslaught. "But I didn't ask what would happen if you eliminated the buffer, just what would happen if you didn't allot any to each task."

While the group considered his response, he continued. "Please, ignore that question for a minute and let me ask another. If I asked you how long it would take for you to drive up the street to Starbucks and get coffee for everyone, how long do you think it would take?"

Then, to illustrate his point, he walked to the board and wrote down time increments, from zero to 60 minutes. "Please, first take a minute and estimate the time in your head. Then, come up and mark an X mark above the time that you think it will most likely take." When they had finished, he walked to the board and quickly drew a graph of the results:

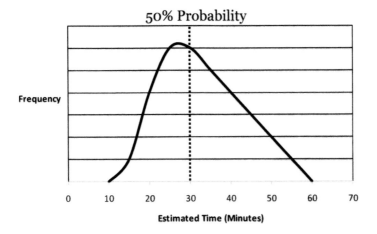

"First, let's determine why some of you estimate 25 minutes and others 60 minutes."

"Well, you never know what the traffic will be like."

"Have you ever seen a Starbucks without six cars in the drive up?" someone else asked.

"Ever heard of calling ahead?" was the comeback.

"Exactly," laughed the Professor. "You are all working from different assumptions. Some of you were thinking that everything could go perfectly, and others were thinking that everything would go wrong. Maybe someone even factored in having to change a flat tire or getting a speeding ticket," which led everyone to single out Danielle, joking with her about her notorious lead foot. "But on average, what did you estimate?" posed the Professor, bringing the group back to the illustration on the board.

"It looks like about 30 minutes," said Dean.

"That seems pretty close," he said drawing a line at 30 minutes and labeling it '50% Probability.' "And what would happen if we used this estimate?"

"We'd get done earlier?" was one uncertain response.

"No, it's more likely that we'd be disappointed half the time," someone else shot back.

"Please explain," the Professor said, interested in hearing his reasoning.

"Well, we'd probably be under half the time and over half the time. That certainly isn't very accurate."

"Why?"

"Because of variation. You know Murphy's law."

"So, if you were trying to create a plan, what could you do about that?"

"Provide some kind of a buffer."

"How much of a buffer?"

"It would take 30 minutes to account for all the variation."

"Yes, but what if we had dozens of such tasks. Would they all use up the buffer?"

"No!" Dean chided, excited again. "Some would be under. Some would be over. We wouldn't need the full buffer. Maybe only half."

"That's pretty close. As a rule of thumb, Goldratt suggests that the buffer be set at half of the project duration," the Professor explained. "But, we still have a problem."

"What's that?" asked Dean.

"If we give each task the task time plus its portion of the buffer, what will happen then?" the Professor asked, challenging the group to find the nugget that would bring it all together."

"The student syndrome," Dean declared, understanding where this was going. "They'll procrastinate, which just uses up the buffer unnecessarily, leaving us without any real protection."

"See, I told you we needed to watch them like a hawk," someone joked.

"Yes," said the Professor, becoming more animated. "But there is a solution. Each task is scheduled at the time it should take based on our 50% probability. But, still we keep the buffer aggregated at the project level. By definition, half of the tasks will finish late, but that's okay because they're late due to variation, not due to procrastination. Then, we track the buffer usage, or penetration, to make sure we aren't using more than we should be as a percentage of completion. You'll see that later."

Building the plan

"So we know how to estimate the tasks and the buffer. Now we need to see how it all fits together in the plan. On the whiteboard, there's a Gantt chart for a very simple project. Can anyone tell me what's wrong with it?" asked the Professor.

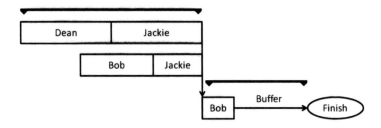

"Well, I'm going to be pretty busy trying to be two places at once," said Jackie.

"That's right," Said professor Y. "How would you fix that?"

"I'd resource level it first," proffered Dean.

"Show us what you mean by that," said the Professor.

"Dean, don't make everything so complicated," teased Jackie. "You just have to delay my start time for the second task until I've finished the first."

"That's what I was going to say," Dean pouted. Then he stepped up to the whiteboard and redrew the second line of tasks. "There," he huffed.

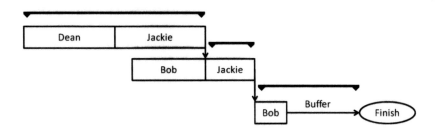

"This is what is known as the critical chain. Just like our process has a bottleneck, or a leverage point, the Critical Chain is the leverage point for our project. We can only shorten the cycle time by shortening the critical chain. But I have another question for you," said the Professor. "How do we know Jackie will be ready when Dean finishes?"

"I don't understand," said Jackie. "I'd just keep an eye on Dean and be ready when he finishes."

"Yeah, like a relay racer," said Dean.

"I'm sure you are sincere, but it never works that way," said the Professor. "What if you were in the middle of running an experiment and he finished a day early? Would you drop whatever you were doing?"

"Well, maybe not drop it exactly. But I'd document whatever I had done and get it to a place where I could safely and efficiently put it aside."

"Why?"

"It wouldn't be very efficient to just stop."

"But that efficiency doesn't create much of a relay hand-off—does it?" he gently chided.

Jackie looked frustrated. "I guess not."

"What you're describing is one of the unintended consequences or undesirable effects of multi-tasking. So, what can you do about it? Anyone?"

"Well, to do a real relay hand off, the other racer must be ready and waiting. The only way to guarantee that is to not be doing anything else," said one of the product managers. "That's not very efficient though, is it?"

"Doesn't that depend what you mean by efficient?" asked Manny.

"Go ahead, Manny," said the Professor. Maggie smiled, sensing the Professor must have seen that Manny was on the right track.

"Well, since the project length is our constraint, we only care about the efficiency of getting the project done. Not the efficiency of any single resource," said Manny. "Besides, the relay is only a metaphor. In our race, there's no rule that says that Jackie can't speed Dean's finish by helping or doing part of his task."

"I always have to carry this guy," grinned Jackie, while poking Dean in the side.

"In your dreams," said Dean.

The Professor rolled his eyes theatrically, "Of course, she could also spend some of that time doing some pre-work to ensure an easy hand off. In either case, what we're talking about here is a resource buffer. Anytime there is a project hand-off, we need a sort of moratorium where the next resource cannot be assigned to work on another project. This protects the critical chain." He proceeded to draw in buffers for both Jackie's first task and Bob's second.

"But there's still one more place we need a buffer. Any ideas where that might be?" he paused, waiting to see if there were any prodigies in the room. "No? Well, what happens if Bob's first task isn't done on time?"

"I'd be delayed in starting my second task," said Jackie. "So, Bob needs to start a little earlier and we need to stick a buffer in there."

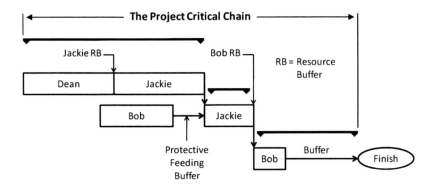

"There you have it. We call that a feeding buffer. Now, one final question and there will be a test afterwards," he winked. "Do these buffers add to the project duration?"

"No, the resource buffers don't because they are just insurance to make sure that Jackie and Bob are available when needed," said Dean. "Neither does the feeding buffer because it's not on the critical path."

"But I don't understand, Professor," said Maggie from the back of the room. "The CCPM plan looks like it takes longer."

"That's a good point. I can assure you that it's not the case, but why do you think that it might look that way?"

"Well, for starters the non-resource leveled plan is an illusion. The CCPM plan is more realistic," said Dean.

"I can buy that, but still, shouldn't it be shorter?" she asked.

"I see the problem, Maggie," offered Manny, "The first plan the Professor drew already had the 50% tasks and the reduced buffer included in it. A traditionally planned project would have been much longer, plus it wouldn't have finished on time anyway because of the resource contentions.

"Very good," said the Professor. "So, you now have the basics for building a critical chain plan. We still need to talk about how to manage using CCPM, but that can wait for another day."

Are we there yet?

That weekend, the Edwards family halted work on the restoration and took a pass on the soccer and Tae-Kwon-Do tournaments. Instead, they packed up the SUV and headed North to Grandma's for the long holiday weekend. Maggie was glad that her mother lived just a few hours away; it gave them an opportunity to see her more often.

Like always, the trip seemed to take longer than the visit itself. It was a busy weekend—her kids enjoyed playing with their cousins, while Maggie and Jeff spent time with her mom and her brother, Sean. After enjoying one of Grandma's big family dinners on the last day, they piled up in the car and headed back home.

It wasn't long before the kids were fast asleep. Maggie was glad it was Jeff's turn to drive—it gave her time to decompress as she watched the scenery fly by. As she stared out the window, Jeff noticed that she was unusually quiet. "What's the matter, Maggs, you look blue. Something big coming up at work this week?" How did he always seem to know?

"Well, not exactly anything big. It's just that I'm halfway into my first hundred days, and it doesn't seem like we've made any progress."

"Maggie, how can you say that? You had an R&D group that was spinning in circles going nowhere and from the sound of it pretty demoralized."

"Well..."

"In just ten weeks you've gotten them focused in on the five most important projects to drive the business forward. That sounds like a lot of progress to me."

"I guess so, when you put it that way. It's just that we're only getting started, and I'm not sure that we're going to be able to move fast enough for Randy."

"Randy?" Jeff rolled his eyes in disgust. "Come on, the guy's never going to be happy. You can't let it bother you. Just do what you think is right and eventually he'll come around."

"I'm not so sure..." Maggie hated the politics and had always tried to avoid situations like this.

"Well, I am," he reassured her as they reached the highway for the last half of the ride home. "Doug is too, or he wouldn't have stuck his neck out, either."

"I hope you're right," said Maggie softly. *I really hope you're right.*

Disappointing Midland

Week 11 – On Wednesday afternoon, Maggie and Manny were discussing project timelines as they walked back from lunch. Naturally, the Energy-Saver system came up. "Maggie, I know we've made a lot of gains, but I've looked at everything possible, and it just doesn't look like we can meet Midland's timing requirement on this project. I've got it down by six months, but that's it."

"Could you use someone from another project or a contractor?" Maggie asked, knowing that it was a desperate question.

"No, it's highly specialized work and would take too long to get someone up to speed on it. I only have one engineer with the right background."

"Alright, Manny, I'll listen to your argument, but we've only taken the first two steps in the improvement process. We really need to do a good scrub and see if there's anything we can do to shorten the critical chain. I'll tell you what. Have the project team in the conference room at 3:00 and we'll go through the issues."

Shortening the chain

"Okay, team, we're here to talk about the Energy-Saver project and what we can do to shorten the project's critical chain," Maggie said, kicking things off. "Dean, as project leader, why don't you give us the rundown?"

Dean took the team through the project plan and the resource loading assumptions. After finishing his summary, he said, "So you can see that even with a CCPM approach which has shaved almost six months, we're still months off of Midland's requirement."

"Thanks for that review, Dean. Now I have a few questions for you."

"Shoot."

"I know you've worked hard on this plan and have made a lot of progress, but Exploit is about working smarter, so I have to ask. Are there other non-bottleneck resources that could be moved to help with some of these tasks?"

"Fair question," Dean said. "In fact, Manny and I have already gone over that. This plan already includes help from the prototype group on tasks 17 and 21."

"Good, I'm glad to hear that." Maggie paused. "Do me a favor and take us all through the individual tasks in just a little more detail and let's see if there are any steps we can eliminate."

Maggie was encouraged when they quickly discovered a step where an entirely new filter housing was being designed. As they reviewed the work breakdown structure and the basic flow diagram, one of the other project engineers said, "Hey, wait a second. I hadn't realized it before, but we could actually eliminate that step entirely."

"Not so fast," said Manny. "The input to that part of the process has to be filtered. As far as I know, there's no similar part available."

"No, I mean we could actually eliminate the filtering step." She grabbed the flow diagram and spread it out on the table. "Over here, this unit would have to have ample capacity, and we could easily cycle some of its already filtered output over to this unit."

"You know, that might work, and after adding the extra few valves and piping, it would still save a few thousand dollars on the system," said Manny.

"Great idea," said Maggie. "Not only have you eliminated a design step for the bottleneck, but you've also simplified the product."

Unfortunately, as they worked the process over the next half hour, they only found a few minor changes they could make, amounting to only a few days off of the critical chain. Maggie didn't let the group know, but she was disappointed. This approach had always yielded more when she had used it in Manufacturing.

Manny could sense her frustration, though, so he stayed behind after everyone had left. "Maybe we could just shorten up the buffer and tell people they have to meet their task commitments," he offered.

"No, Manny. If everything goes perfectly and we don't have to use the buffer, that will be great, but we never want to start thinking that way. Those tasks estimates are just that—estimates. Start calling them commitments and everyone will start padding them in the first place. Then we'll be no better off than when we started."

Elsewhere in Barrister's world

Chuck Jacobsen, the VP and General Manager for the Barrister Industries, Electron Instrument division, listened as one of his project teams explained the automated lab sampler they were working on and their launch plans. The more he heard, the more unsettled he became.

"The new HPZ 212 series will include our patented Intelli-Chip technology designed to set a new standard in its class." He had listened to the droning discussion for 15 minutes and still wasn't sure what it was that the product was going to do differently than the competition. "Additionally, all of the electrical components will meet the latest standards for EMF shielding and efficiency."

Chuck was not happy with what he was hearing. After sitting through a one-hour new product team presentation, he still wasn't sure that the team had the right target. He'd heard many buzzwords, but he was afraid that even the product managers were in love with the technology and were marketing features, instead of benefits. He'd asked a few questions, but still wasn't given a clear benefit. He decided to try a more direct approach. "Can you tell me what the key product benefits are?"

"Certainly, the frame will be built with superlight manganese alloy and the sampler controls will be state of the art."

"And what benefit is that going to give the customer?"

"They'll get class-leading performance and exceptional value."

Chuck felt like they were going in circles. Rather than beating up the team, he decided to drop it there. But he knew he needed to do something or this new product, and possibly those following it, would break the string of new product wins that had started this division.

Chuck asked his direct reports to stay behind when the meeting was over. Chantal Watts was the R&D manager, and Mick McClain was the director of product marketing. With the room to themselves, Chuck asked, "What was that?"

"Well, the marketing needs a little work," agreed Mick. "But it's early."

"A little work? I'm not sure any amount of marketing can save a product when we aren't even sure what real problems it's solving," said Chuck. Turning to Chantal, he added, "And I'm not sure your engineers haven't just included every gee whiz feature possible without giving consideration to cost."

"Well, I'll admit the cost is looking a little high right now, but if we can get to 1,000 units, the prices come down to an acceptable margin."

"Look, the two of you need to get a handle on this before it's too late." Chuck was not going to let the first new product since he'd become GM fail. "I'm out of the office quite a bit over the next two weeks. In the meantime, I want you to put your heads together and give me an overview of how you are going to get this product back on track."

New terrain

Week 12 - Sitting in a resort conference center outside of Atlanta, Maggie looked out the window at Stone Mountain and the beautiful scenery. She wished she could enjoy it more, but she was feeling guilty that she wasn't at DFT where she could get something done. Besides that, being here meant that she had to leave Jeff with the kids—not that he couldn't handle them, but it never quite felt fair to her.

She was here for one of Doug's quarterly management reviews. She wished that Doug would hold these meetings at a conference center in town, but she also knew that with so many of the divisions spread across the country, the expense difference was minimal. Plus, it helped keep folks focused for the two days.

As they finished the first day's sessions, part of the group headed out to try and get at least nine holes in while there were still a few hours of daylight. She found herself walking out alongside Chuck Jacobsen from the Electron Instrument division, who asked, "Don't you play golf, Maggie?"

"I do, but I noticed they had some mountain bikes available and thought I might get in a ride to burn off that lunch."

"You're a biker, then?" he replied.

"Not really, but honestly, I could really use more of a workout than resort cart golf provides."

"I can understand that," he laughed. "Do you mind if I join you?"

"That would be great. It's probably not a bad idea to have someone along to call Flight for Life in case I ride off of a cliff or something," she joked. "Why don't we meet in the lobby in fifteen?"

After an hour of hard off-road riding, they came across a fire road that looked like it would take them back to the resort. It was a little longer that way, but neither of them were teenagers any longer and finishing up with something a little less bumpy was appealing.

As they pedaled their way back, Chuck said, "So Maggie, Doug tells me that DFT is doing some pretty exciting things in new products. That sure wasn't the case in the review session last year. Honestly, if Roger wasn't one of Randy's hunting buddies...Well, let's just say no one was surprised by the change."

Maggie didn't want to get into a discussion about Roger, so she shifted gears, picked up her pace to get a little bit out front and said, "The team has really pulled together. I'm very proud of them."

Chuck matched her speed and continued, "Well, I know that the change in leadership had something to do with it. What's your secret?"

"Just good people and hard work," she said feeling a little embarrassed.

"Maggie, I'm serious," he called, falling behind. "I really want to know. While Electron has done well since start up, I'm concerned about future new products. What have you done to turn things around?"

"You really want to know?" she said looking back over her shoulder.

"Yes," he puffed.

Maggie slowed down to let Chuck catch up. "I'm afraid you're going to be disappointed because we haven't done anything fancy."

"Who needs fancy? We need improvements."

As they pedaled alongside each other, Maggie spent the next 30 minutes explaining the changes they had made at DFT.

"Wow, I know we've used TOC in manufacturing, but I didn't realize it had application to new product innovation."

"Neither did I, Chuck. What do they say about necessity being the mother of invention?"

"You know, I've talked with a few of the big consulting firms and they wanted to come in and implement a full-blown stage-gate process. I've given it some thought, but that's a lot of money and added complexity with no guarantee it will deliver more new products—certainly not in the near term. I really like the simplicity of what you've done. It's an interesting alternative."

"I need to warn you though, Chuck, I'm not sure any of the things we've done would help you. I suspect your journey would be very different than ours," she said.

"I don't understand. Why can't we go to school on what you've done?"

"This is a framework for making high-leverage improvements. The first problem is that your leverage point or bottleneck could be in a different place. The second is that there are probably different issues constraining you from getting everything out of it."

"That makes sense. It sounds like it really depends on where your bottleneck is."

"That's right. Chuck. What's your biggest concern? Where do you see your bottleneck?"

"Well, my engineering group always feels it doesn't have enough resources, but frankly, half of what they're working on is questionable. It's almost filler work. We created the market category we're in, and somewhere along the way, we've lost our market direction. Honestly, Maggie, I think we've gotten lazy. We're enamored with our own ideas and technology, rather than taking the time to understand what the market needs."

"Those are symptoms. What is the bottleneck that's causing them?"

"I see your point. I'd have to say it's finding good new ideas—problems that customers will pay us to solve."

As they turned onto the main road leading to the resort, she asked, "Chuck, do you have any idea what operating policies might be allowing that to happen?"

"No, Maggie, not offhand."

"Well, honestly, Chuck, you shouldn't answer that question by yourself, anyway," she said, proceeding to explain to him the three steps of what to change, what to change to and how to cause the change.

"No question, I need to get Chantal and Mick's teams involved in developing a solution if we're going to make any progress."

"Not just them, it really takes involvement from across the organization because they are all part of the solution. You know, Chuck, there's a professor at the local technical college that has consulted with us on TOC for many years. Lately, he's been helping me figure out how to apply it to the new product development process."

"Do you think he could help us?"

"I'd be happy to talk with him," she said as they pulled up to the front entrance and dismounted, their legs wobbly from the hard ride. "We created a new product project assessment tool, and my marketing lead, Danielle Espinosa, has been dying to develop it further into a tool that she calls *Customer Value Lens*. It's a tool to focus resources on identifying unmet market needs and the value that you can create. Perhaps she and our consultant, Professor Yulinski, could spend a day or two with your team. But I am guarding Danielle's time carefully."

Chuck jumped at the offer and said he was sure he could come up with the consulting fees for the Professor and cover the travel for both of them. Maggie wondered whether they really had the time for Danielle to make the trip, but she wanted to try and be a good team player. And, truth be told, Danielle had really piqued her curiosity about the Customer Value Lens. She was interested in seeing how this would work in practice.

Focusing the customer value lens

Week 14 - Danielle and the Professor traveled to Electron Industries, where they met with Chuck Jacobsen and his team to see how they could help with their new product issues. The Professor started with a thorough review of the TOC concepts that DFT was using to drive new product improvement. Then he used his usual inquisitive approach and confirmed that, indeed, the bottleneck was where Chuck had thought it was. Electron was not finding enough new opportunities.

"Now, given that we know where the bottleneck is, let's talk about what might be constraining it. What is preventing you from finding more good opportunities?"

After 20 minutes of brainstorming followed by evaluation, the group quickly narrowed their list to the big issues.

- Haphazard approach to finding opportunities
- Limited customer contact
- Not spending enough time with users
- Difficulties obtaining economic data
- End user is not our direct customer

At that point, he turned things over to Danielle. "As the professor covered, TOC measures improvement in three ways: increasing throughput, decreasing investment, and reducing operating expenses. While these measures are important to Electron, they're also important to your customers. Who's heard the old saying '*Customers buy benefits, not features?*'"

Nodding their heads, everyone acknowledged that they had heard the phrase. "Improvements in these three areas, or reducing the risks surrounding them, are pretty much the only reason that an industrial or business-to-business company will adopt a new product, and they form the basis for something I call the Customer Value Lens. It's all about finding the big benefit that your customer will pay for." With that, she walked to the board and wrote:

- What jobs are customers trying to get done that are costly, inconvenient, time-consuming, unpleasant, or even dangerous today?
- What is the big benefit that you could offer relative to the market's unmet needs?
- How would those benefits create value for the customer in terms of:
  - Increased throughput?
  - Reduced investment?
  - Decreased operating expenditures?
  - Reduced risk?
- How would you share in the value created?
- What new competitive dimension have you found?
- What would your distinct advantage be relative to the current competitive alternatives:
  - Higher performance?
  - Lower cost?
  - More convenient?
  - Easier to purchase?

"I hope you can see that this lens has value in helping you find the problems you can solve and then later in marketing the solutions that you develop." Danielle then readied herself for what she expected to be the hard part, holding up copies of one of Electrons latest new product plans that Chuck had shared with her. "Now, I did spend some time reviewing your plan for HPZ 212 series ahead of this session. Sorry to be so blunt, but I don't see too many of these elements in your plan."

To her surprise, there wasn't any argument. Just downfallen looks as if they'd been struggling with this for quite a while. Chuck seemed to be a nice enough guy, but there was a lot at stake in their quest to get this second big product line for Electron right, especially with the cash flow issues facing Barrister. She wouldn't have been surprised if he'd been pretty hard on them about this. "And if you don't start a project off with these elements, it has to be pretty difficult to market it when the time comes. How many of these elements are covered in your marketing?"

"Only some," admitted Mick McClain, the head of product marketing. "I have to admit that we haven't been very comprehensive. I think we talk more about technical features than true benefits."

"How would you suggest we change that?" asked Chuck.

"Well, you can only answer these questions by getting people out of the lab and their offices and into the field, where customers are actually experiencing these problems."

"But we have technical people on staff that used to work in our target markets," protested Chantal.

"That's a good start, and their network can help open doors, but markets change. One of your primary markets is genomics. That's such a fast moving area that the market knowledge you had two years ago isn't necessarily relevant today. But it's more than that. By relying on one person's knowledge or recollection, you're limiting yourself solely to the problems they were aware of. It's important to get out and open your horizons."

"But that's what product marketing managers are supposed to do," replied Chantal.

"Well, yes, but sending them out by themselves means you're only getting one perspective. Plus, you're relying on them to translate, when they may not have the same ability to see technical problems or solutions—even if they're technically trained."

"I can see that," said Chantal. "Maybe we really do need to start doing this cross-functionally if we want to make those benefits part of our program."

Danielle smiled at Chantal's realization. "Well then, let's get to work on making those benefits a key part of your new product ideation."

For a downloadable version of
The Customer Value Lens go to
www.SimplifyingInnovation.com/extras

Priority logic

Week 15 - "Maggie, I'm afraid I've got more bad news." Manny had cornered her as she hurried back to her office for a Monday review meeting with Doug. Because of the proximity, Barrister's CEO liked to escape when he could.

"I'm meeting with Doug in a few minutes, but let me ask you this? Is it about the Energy-Saver system?"

"No...But now that I think about it, we might be able to shave some time off the project."

She had a confused look.

"Go. I'll explain later, but don't promise anything."

Maggie agreed. Doug arrived at her office a few minutes later, and they quickly worked their way down the review agenda. As they were wrapping up the discussion, Doug said, "Maggie, there's one more thing. I really hate to bring it up, but I don't have much choice. Randy is all over my back because he's heard that you're not working on the highest return projects. I told him that couldn't be the case and to leave it to me to talk with you."

"Well, I appreciate you running defense for me, but I really don't have any problem talking with him. In any event, I can guess what he's hearing and who he's hearing it from."

"Roger, huh?" Doug seemed to know, too.

"He doesn't want to set any priorities because he's still pushing to get everything done at the same time. I told you that Randy had stopped by to express his displeasure about that a few weeks ago. I was hoping I had alleviated his concerns. But this is about something else. Do you remember how I told you we were going to set priorities?"

"Sure, based on the NPV for the project."

"That's right, assuming that all the projects fit equally well with our competencies and our strategy, but more specifically we're prioritizing based on the NPV per unit of bottleneck, or design engineering time. One of the things I learned at Terra-Grafix was to prioritize based on the return or the throughput per unit of bottleneck time. If you recall, the assembly step was our bottleneck there. We had two different systems we could build, and while it's not rocket science, moving from producing 8 units per unit of assembly time with a return of $5,000 each to producing 5 with a return of $10,000 each increased our throughput by 25%. We used this very effectively to squeeze the items with the highest throughput per unit of bottleneck time through the operation first.

"Sure, I remember that's when we saw that big bump in cash flow in our first year of TOC.

"Exactly. Hang on just a second, let me bring up a file to show you something," she said, while walking over to her computer and pulling up a chart with twelve projects on it. She highlighted two of the projects and printed out copies for each of them.

	HPGR Project	SDFT Project
Projected Return	$1,500,000	$1,000,000
Development Time (hours)	2,000	1,750
Return per Development Hour	750	571
Ranking	1	2
Design Engineering Time (hours)	1,500	750
Return per Bottleneck Hour	1,000	1,333
Ranking	2	1

Now if you'll take a look at these two projects, you'll see that the *HPGR* project has a higher NPV and a higher ROI, and with a similar total development time, it has a higher return per total development hours."

"Okay. But I see here that it takes almost twice the design engineering hours, so the return per hour of bottleneck time is about 50% better for the *SDFT* project."

"Exactly. Since we have ample capacity in the other areas and are limited by design engineering capacity, we'll get more contribution from the lower NPV project that uses less design engineering hours."

"Thanks for your help with that, Maggie." Holding up the printout, he asked, "Can I keep this? I'd like to review it with Randy and you can consider the case closed."

Step 3 – Subordinate

Only halfway there

As soon as Doug left, Maggie went to Manny's office to find out what the bad news was and how it could possibly help them save the situation with Midland.

"Okay, Manny, what's up?"

"Well, first of all, the bad news. The project team for the ultra compact system has completed their initial feasibility testing, and it's a no go."

"What do you mean? Why not?"

"Do you remember that one of the critical assumptions on the project was that we could deliver a larger payload by operating in a different flow regime?"

"Sure."

"Well, it turned out to be true, but the effect was less than half of what we had predicted. The extra performance doesn't justify the cost."

"Manny, I want you to do something for me." She went on, asking him to have Nancy put together a small recognition meeting that afternoon. Nothing fancy. Just the project team, Danielle, Manny and herself for snacks in the cafeteria.

"I'm not sure I understand. The feasibility testing failed."

"Manny, had we not added the feasibility testing step, this project would have gone on much longer, consuming constrained resources. The system's working!"

"I understand that, but I hadn't thought about celebrating."

"We're not celebrating, just recognizing and reinforcing that they did the right thing," she said. "We need to make sure that teams know that we don't shoot the messenger when they are brave enough to kill a project—especially when they save critical resources by killing it before design engineering starts."

"I'll work with Nancy to arrange something," he said.

"So, how does this impact the Energy-Saver project?"

"Oh, yes—the good news. One of the engineers who was on that project has some of the expertise we need. He's not a direct plug in, but I still estimate it could help us shave off half of the delay."

Maggie was glad to hear it. One of the advantages that she had expected from pipelining projects was better visibility into the workings of each project and they were getting it. "Thanks, Manny. That is good news," said Maggie. Unfortunately, that wasn't going to be enough to satisfy Midland.

She didn't have to say anything more. They both knew they still had to find a way to close the rest of the gap. Maggie had a nagging feeling that she had forgotten something. "Manny, before you go, what is the ultra compact project team going to get started on next?"

Momentarily confused, he stepped back in and replied, "Well, the next highest priority, I suppose."

"So, is that project all teed up and ready to go for the Design Group?"

"Oh sh..." he caught himself. "I'm really sorry. I didn't expect that we would need to have another project ready this soon."

"No, I'm not blaming you," she said, trying to alleviate his concern. "We all missed this in exploit. Look, we're still in start-up mode. The important thing is that we need to take steps to make sure the non-bottleneck resources establish and maintain a protective buffer of two or three projects always waiting and ready to go to Design."

"Sure, as soon as one project is released to Design, they need to begin working on getting another ready. Well, it's good that we caught it now, so we can get it fixed. I'll meet with my team and get right on it. You know, I can see that conducting traffic is going to be important."

The train has left the station

Later that week - Danielle and Manny were waiting in Maggie's office when she returned from lunch. "Are we ever glad to see you," said Danielle.

"Okay. You both look a little anxious. What's the problem?"

"We were visiting a few big customers to determine if they were going to be good candidates for the Energy-Saver filtration system, when they mentioned that our competitors are going to launch a system with variable speed pumps."

"So, what's the problem?"

"Well, Roger sort of promised that ours would, too. Unfortunately, that will add at least three months to our schedule because we'll have to go back to the engineering design group. It's going to add cost, too, not only for the extra design work, but for the finished system, as well."

Wondering if Roger was doing this just to cause her to fail, Maggie asked, "How will our system perform if we don't have that feature?"

"It only adds another percentage point or so of efficiency. It's nothing to sneeze at, but we far outstrip that with our laminar cascading filter bed design. And that's something the competition doesn't have. Of course, it doesn't take much added efficiency to pay for the variable drives, so the changes would add value for some percentage of the users."

"This doesn't sound like a fatal flaw that we missed in the planning. Does it?"

"I don't think so," said Danielle. "It's an improvement that some customers will pay for, but certainly not a fatal flaw."

"Sounds like Bright Shiny Object Disease."

Manny hadn't been part of the earlier conversation, so Danielle explained, "BSOD in this form would be someone seeing new features in the marketplace and wanting to add them to development projects that are already in design. The bright shiny object is a distraction from the agreed plan."

"That's right," said Maggie. "Those features might actually be a good idea if design hasn't already been started, but if it has, adding them could be very disruptive to the process. Honestly, it's something that I thought was causing a lot of delays and finger pointing when I came on board. Almost every project had design changes made to add new features after design had already begun and sometimes after it had finished."

"We definitely shouldn't disrupt the project," said Manny, "But how can we say no at this point?"

"I know it's awful to continually bring up experiences from TerraGrafix, but we had a similar situation that we might be able to learn from. One of our bigger workstation systems was highly customizable, and our salespeople sold the heck out of that fact. Unfortunately, they were still trying to upsell customized features, sometimes even after we had started building the system in manufacturing. Needless to say, the disruption was killing us. We finally solved the issue by instituting a policy that included a much higher price, lower sales commission, and two months longer lead time if options were added after the contract was signed."

"Sales must have screamed about that!"

"A little, but not nearly as much as if we had forbidden changes altogether. We still allowed changes. It was just that there was a cost and lead-time associated with them. After that, I can only remember one or two requests for so-called change orders. When most customers saw the added cost, they said, 'Oh, don't worry about it. It's not that important.' And revenue didn't drop off at all. Sales knew their only chance at the bigger commission was to get the optional added features in the original contract, so they worked to include them earlier. Also, since this helped us cut lead times and reduce delays, we eventually started getting more orders because competitors couldn't deliver as quickly."

Manny jumped in. "I think that's a great way to handle it. We should always freeze the requirements for any product once it enters design. We don't say no, but we make variable speed an option in Version 2 of the product, which we don't announce until several months after the base model comes out."

"Alright, I was probably one of the biggest violators as far as changes, but this makes a lot of sense," laughed Danielle. "Frankly, with what I've learned in the last few months, I can't think of a better way to make sure that new products never finish than to continually add features after we've started development." Maggie was glad to see Danielle buying in.

"Sort of like calling back the train after it's left the station just because another passenger wants to board," observed Manny.[24]

"Great, so let's tell Roger and Sales what our new features policy is and that the feature he promised will be available in Version 2," said Maggie, realizing how lucky they were that he hadn't made this promise to Midland.

Bang goes the drum

Week 16 - It had been more than a month since the DFT teams had started using critical chain, so the Professor returned to review progress and make the final connection to the DFT innovation process. "Let's review what we accomplished last time," said the Professor:

- "We estimated our individual tasks at 50% probability by cutting the total task time in half,
- We cut the total amount of buffer in half and aggregated it at the project level to prevent it from being wasted on procrastination,
- We leveled out the resources so that no one is scheduled to do two tasks at the same time,
- We added resource buffers to ensure relay-like hand offs,
- And we added feeding buffers to protect the critical chain from delays."

"But there's still one more constraint we need to be concerned about." The professor paused and looked around the room, but no one seemed to know where he was headed. "What happens when you try to share resources across multiple projects, even if they only work on one task at a time?"

"Someone usually ends up waiting. But can't we use the estimates to schedule them better?" asked Dean.

"Dean, by definition, our task estimates will be wrong half the time, and we will use some amount of the buffer that is impossible to predict. That makes it rather impractical to schedule each and every task ahead of time."

"Sounds like we're screwed," laughed Jackie. "But somehow, I'm betting you have another card up your sleeve."

"It does sound rather grim, doesn't it?" The Professor smiled. "But you're right about the other card."

Manny had been quiet up to this point, but suddenly, a look of inspiration crossed his face, and he said "Subordinate."

"Go on," said the Professor with a look of satisfaction.

"Well, the third step in improvement is subordinating resources to the pace of the bottleneck—to its drumbeat. We've decided that to best exploit our bottleneck, Design Engineering, we are limiting their workload to no more than five tasks across the entire group. So as soon as they finish a task, we recently decided that there should be another one waiting and ready to get started. It's like the inventory safety buffer that you keep in front of a manufacturing bottleneck."

"That's fine for the Design Engineering group, but it doesn't help make efficient use of everyone else's time," interjected Dean.

"We're not necessarily concerned about the efficiency of non-bottleneck resources," said Manny, "but, since our goal is to finish projects on time or early, we do need to share those resources in a manner that helps each project finish as soon as possible."

"The Professor just said it was impractical to schedule each and every task ahead of time," Dean wasn't going to let go of this one.

"That's right, but I'm suggesting that we need a dynamic approach that ensures that they know what to start on next after they've finished another."

"So, how would you decide what they should work on?" asked the Professor. "What is your drumbeat?"

"Again," said Manny, "our goal is to get each project done as early a possible, so I would say that the highest priority should be the project that is running the furthest behind. Maybe we can prioritize by the percentage of buffer consumed."

"Indeed," said the Professor, "and it's called buffer penetration. Let's make sure we cover that later.[23]"

Preparation is not optional

Week 18 – Exploit at Electron Instruments

During the session at Electron, it was clear that part of the issue was that they didn't have a systematic approach to finding unmet customer needs. A key element of exploiting their new product development constraint was conducting better customer visits. So, Danielle agreed to help them conduct a few customer interviews to see why they were struggling. Her first clue that they had a bigger problem came when several folks declined an invitation to meet for the purpose of developing the interview guide for the customer visits. Even worse, they'd waited until the morning of the meeting to do so. The responses to her email asking why they had declined indicated they thought it would be easier to meet the morning of the visit and the team could talk about it then.

Great, we can all just meet at a greasy spoon up the street from the customer and scribble some questions on the back of a takeout menu, she fumed to herself. *No wonder these guys are struggling. They're just going into customer interviews and winging it. No planning, no forethought about why they are going in, and no idea what they want to learn from the call!*

After calming herself down, she made a quick phone call to Mick and Chantal to explain the situation. She hoped she didn't have to convince them. She was happy when Mick replied, "Yeah, I saw the decline notices, and I'm really sorry about that. I tried to make it clear that the status quo wasn't going to cut it anymore. Evidently, I'm going to need to have a remedial chat with a few folks. I'll get right on it, and you can count on everyone being on that phone conference this afternoon, Danielle, and they will be prepared." The message evidently got across because everyone phoned in for the conference and on time.

Tasks will be late

Maggie and Manny were leaving Barrister corporate head-quarters after completing another progress meeting with Doug. As they headed toward the elevators, Randy Barrister stuck his head out of his office door. "Maggie, could I see you for a few minutes?"

She was surprised to run into him since he had a reputation for being an absentee leader. Maybe he was taking things more seriously since the cash flow problem had come to a head. "Sure, Randy. I've got a few minutes. Manny, do you mind waiting?" she asked before steering toward Randy's office.

"I'm glad I caught you. I'm very concerned about some things I've been hearing," he said as they took a seat in the big leather chairs in his office.

"I'm sorry to hear that, Randy. I thought things were going rather well. What is it that's concerning you?" she asked, thinking that this guy seemed to be looking for things to be unhappy about. Didn't he have any acquisition issues to deal with or some yacht club event to attend?

"I understand that a number of projects are falling behind and that nothing is being done about it."

"I don't mean to be evasive, but could you be more specific? As you know, we're running five projects, and I get regular updates on all of them. All of them are still projected to finish on time and on budget."

"Well, I hear that some of the tasks in your redundant filtration project are already behind schedule."

"Well, that's right, Randy. But..."

"Maggie, I told you the last time we spoke that I was concerned we had made a mistake...

"Randy, please let me explain," she continued, waving him off. "Better yet, let me get Manny in here. He can explain it even better than I can." She didn't mean to drag Manny into the lion's den, but wanted Randy to see their solidarity.

After Maggie explained Randy's concern, Manny knew immediately what the issue was. "Yes, that's right, Randy, the design phase on that project was late, but that's to be expected.

"I expect just the opposite! I thought we were hiring you to fix that, and now you tell me it's normal?" Maggie knew this tirade was directed at her.

"Randy, it's normal because we estimate each individual task at the amount of time it will take to finish with 50% probability," Manny paused to see if he was listening.

"Go ahead."

"Then, we add a safety buffer for the entire project. Using that approach reduces project times by as much as 50%.

"I still don't see how any competent project manager can accept a project being that late," demanded Randy.

"Randy, I understand your concern," said Manny. "I have to admit, that was one of the hardest things about Critical Chain Project Management for me to accept, too. The project isn't late. The task is late. In fact, by definition half of the tasks will be late."

"Still..."

Maggie was glad to see that Manny was keeping his cool and that his confidence in this approach was building as he forged ahead, "But half will be on time or early unless there is some unexpected delay, and that's why we have the safety buffer. The important thing is that the team monitors the project buffer to make sure that it's not being used up too quickly and that the cumulative result will be an on time finish. If not, then we take action."

"And how do you keep track of that? What steps are you taking to stay on top of this?"

"I'm glad you asked. Can I use your whiteboard?" Randy motioned for him to proceed, and he started drawing a simple graph. "This is a project management tool that we've started using to track all the projects now that we have a reasonable number and better plans,"[23] he said, sketching a graph showing percentage completion of critical chain tasks vs. percentage buffer used. "The upper line here is our warning line. If a project ends up above the line, like Project C here, then we have to take action to recover some of the buffer. This bottom line here is our watch line. Anything below it is in excellent shape. So, we don't need to do anything other than continue to watch project A."

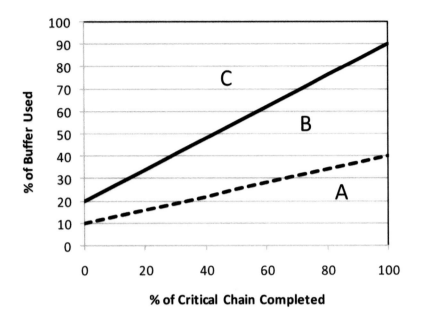

"What do you do here in between them?"

"That's the planning zone. If a project starts to edge above the watch line, like project B, then we need to begin contingency planning to recover part of the buffer."

"Why can't you hire better project managers and bring projects in sooner to match the lower line?"

"Each task on that lower line only has a 50% probability of finishing on time. Multiply those probabilities and the chances become very low." He continued to explain how the student syndrome and Parkinson's law affected projects.

"But if people commit to getting a task done in a certain amount of time, they should."

"We don't ask for commitments. We ask for estimates because we know full well that Murphy will strike. And the buffer is there to account for that."

"Yes, but just think how much time you could save by not using all of the buffer."

Maggie had to stop this. "Randy, believe me, we're always looking for ways to finish projects sooner. It's a key metric for new product development. But we've already cut out 50% of the buffer versus what we used to allow, and by managing it in this way, we're actually staying on schedule. If you combine that with actually finishing on time, were going to complete projects in half the time—even less if you include all of the projects which were never going to finish," said Maggie.

"Okay," he said stabbing his finger into the desktop, "But I'm going to be watching these projects closely. We can't afford to miss. I'd like to start seeing this chart for all your projects on a regular basis."

Maggie couldn't believe this guy. He'd been distracted by acquisitions and had paid no attention as DFT's innovation ground to a halt. Now they were making real progress toward cutting the average project cycle time in half, and he wanted to help by micro managing. Oh well, there was nothing she could do about him. She just needed to keep her head down and make sure the team stayed focused on delivering the improvements she had promised. Of course, she had to do even better than that if they were going to hang on to Midland Industries.

Hijacked

Danielle felt confident as she walked into Patterson Pharma with the Electron team. They were prepared to make the first of their customer interview calls with George Pappas and Dave Jackson, the general manger and the lab manager. Whenever possible, it was important to get to the managers and engineers experiencing the problems—dealing with the pain and the operating costs, rather than purchasing staff who tend to focus on acquisition costs.

Danielle was glad that the Electron folks had followed her suggestion about having both technical and commercial folks involved. She also appreciated that the calls were here in town so that she didn't have to travel. Sanjay Giri, one of Electron's product managers, was assigned to lead the interview and ask the questions. Bob Ryan, the R&D engineer, was assigned to ask follow-up or clarification questions, and Danielle would be the scribe. They were accompanied by Patricia Davis, the distributor's representative for the account.

As hard as it was for Danielle, she reminded everyone about silencing their phones as they walked into the lobby. "You can't stop the customer from taking a call, but you can make sure you never interrupt a visit."

Danielle had asked them to schedule 60 to 90 minute sessions with 30 minutes of reserve time, just in case the customer wanted to continue. It was important to have enough time for a thorough discussion but not to go overboard and wear out your welcome. The first ten minutes of the call went reasonably well. Sanjay was asking the open-ended questions they had put together beforehand and was moving the conversation along, when suddenly Danielle was shocked to hear Patricia answer one of the questions. Patricia certainly knew the account very well, but they were here to find out what George and Dave thought. She could have kicked herself, immediately realizing that Patricia had not understood what she meant when she said that customer visits were not sales calls. She'd have to make that abundantly clear before the next call.

Then it went from bad to worse. Bob had asked a great clarifying question, and George responded by going into some depth about one of the problems they had in getting samples prepped uniformly. So far so good, but then Patricia pounced. She was a natural salesperson and the customer had laid out a problem in front of her. She was trained to find opportunities and present solutions. How could Danielle blame her?

From that point on, Danielle noticed that the tone of the conversation had changed. Even after Sanjay got them back to the original line of questions, the information was harder to get. While Patricia was clearly respected, there was now some information George and Dave just weren't comfortable sharing. They had lost control, and it had become a sales call. Danielle nodded discretely at Sanjay. *Too late—just leave it alone.*

Walking out to the parking lot, Danielle decided that it was best to address the situation before the next call, so she asked Patricia to ride with her. As they pulled out of the parking lot, Patricia asked how she thought the call had gone.

Here goes nothing, thought Danielle, knowing they still had three more visits to make. "While I'm sure that went well as far as a sales call, I have to take the blame for letting it become a sales call instead of a customer visit." She then went on to explain what she had seen and how it had prevented them from getting the information they needed.

"Danielle, I am so sorry. I didn't even realize that I was doing it. I guess it's just part of my DNA. You know an autonomic response like breathing," she laughed.

"I'll be honest. I'd be a little reluctant to have you on this next call, but I know you're a professional. You didn't do it on purpose. I just need you to promise me that you will limit your participation to introducing everyone and observing."

"You won't hear a peep out of me."

When they reached the destination for their next call, the group stopped at a small diner and used the time they had scheduled between the calls to debrief on what they had learned. Preparing the interview guide beforehand had given them plenty of time to discuss and write down the important learning from each call and where appropriate to incorporate them into the questions for the next call.

Win-Win

The second call that day was at Hamilton Labs, where they were meeting with one of the lead researchers, Dr. Lim. Patricia kept her promise during the call, and Sanjay and Bob did a great job moving the interview along through the first two thirds, which were devoted to uncovering problems and issues and understanding them better. Things looked very promising when they uncovered the same issue that had come up on the previous call.

That's when they hit another snag. After Dr. Lim shared a significant amount of information about the problem she was seeing, Sanjay asked, "How is that affecting you?"

"I'm afraid it's raising our costs significantly and reducing our output."

For clarification, Bob asked, "And how is that impacting your bottom line? Have you quantified that yet?"

Dr. Lim stopped abruptly, her radar going up. "I'm not sure that it's appropriate for me to share that with a supplier. Wouldn't you just use that information to get a higher price?"

Danielle knew that neither Sanjay nor Bob were prepared to answer that question, and she could see that Patricia was becoming very uncomfortable. She touched Patricia gently on the arm, letting her know that she wanted to handle this.

"Dr. Lim," said Danielle, "I admit we are trying to understand how much value there would be in solving your problem. But that's only because it has to be a win for both of our companies. We hope to find the problems that are causing you the most frustration and then develop new products or services that alleviate them. But before our parent company, Barrister Industries, will allow us to invest in any new development, we need to know two things. One, that a solution will provide you with significant value, and two, that there is enough value in that solution to provide a reasonable return on the sizeable investment that we'll have to make to bring the product to market. Again, it has to be a win for both of us. Does that sound fair to you?"

Danielle could see that she was warming to the approach. "It does," replied Dr. Lim. "I spent a few years in purchasing, and sometimes it's easy to fall into the habit of thinking win-lose." She then went on to lay out the problem and its cost impact in more detail in what turned out to be the best of the three calls that day. She even agreed to let several of the Electron project team members come back and spend a day observing their labs in operation. That would be tremendously helpful in identifying the jobs that end users like Hamilton were trying to get done.

As they drove to the next customer, Sanjay piped up from the backseat, quickly running through the numbers on his laptop. He was so excited that he was almost vibrating. "Danielle, I really liked the way that you explained why it was in Hamilton's best interest to share that information. Sounds like our automated sampler could easily help them double their business in genetic testing. With their work backlog, that's worth millions in profits to Hamilton at this location alone. If my calculations are correct, a 12-month payback for them would still provide us with a 65% margin vs. the 30% we normally have to fight to get. Now, that's win-win. Plus, this has to be an issue for an entire segment of customers."

"You know, Sanjay, we once worked all the way through prototyping on a new product, only to find out the target market couldn't afford our solution," said Danielle, "Being able to get this kind of understanding about the customer's value upfront is invaluable to fully exploiting your innovation bottleneck." She couldn't help smiling as they pulled into the parking lot of her favorite sandwich shop. The *Customer Value Lens* was working even better than she had expected.

Not adding up

As another week ended, Maggie felt like her challenge had turned into an onslaught. When she arrived home Friday afternoon, the twins presented her with parent/teacher conference notices and their report cards. They'd always done well in school, but their math grades had started to drop in this period. As fourth graders, they were encountering more difficult word problems, and Maggie knew that what they needed was just a little help understanding them.

She sat the boys down at the kitchen table where they could spread their books and papers out and she could tutor them while she started getting dinner ready. While Luke and Leo focused on improving their math, Maggie contemplated the improvements she knew she had to make.

Not only had she received parent/teacher conference notices, but earlier that day, Doug had informed her that Randy had requested a 100-day review session with the executive board in just a few weeks. While Doug was quick to brush it off, Randy had made it clear that he was dissatisfied. On top of that, Roger had no doubt been giving him an earful about the Midland situation. Either way, it didn't give her much time.

In a way, Maggie knew that Randy was right—not about how to manage innovation projects or even the portfolio, but about the results. She appreciated the fact that Doug was standing up for the progress they were making, but still it wasn't enough for her, either. Maggie expected to see the results come faster with TOC, just like it had when they had used it in manufacturing. She doubted that the executive board would give her a passing grade for her results so far.

Things were moving in the right direction, but there were still too many small setbacks. Had she been wrong about the bottleneck or about the policies that were constraining it? She really didn't think so. But then why weren't things clicking the way she had hoped? Like Luke and Leo, she wished she could sit down with someone who could help her understand the impediments to their progress, but the Professor was out of town for the next few weeks. Her review session with Randy was only three weeks away, so she was going to have to count on Manny and Danielle. Maybe they could spot something she wasn't seeing, because right now she could really sympathize with her boys—something just wasn't adding up.

Short fuse

Week 19 – A quick glance at the display on her ringing cell phone told Maggie the call was from someone at Barrister Industries. *That's unusual*, she thought, *who would call her cell phone, rather than her office number?*

"Hello Maggie, this is Randy Barrister's assistant, Tiffany. Sorry to bother you, but Randy's schedule has changed, and he needs to reschedule next week's review session."

"No problem," said Maggie. It looked like she had caught a break and would have a little more time to address the new products issue before she had to present to the board.

"Randy would like you to present Thursday."

"Thursday? This Thursday?" Unfortunately, this was a command performance, "Well, I guess we'll have to make it work, won't we?"

"I'm really sorry to do this to you."

"No, it's not your fault." She wouldn't be surprised to find out that Randy had done this on purpose. He was becoming less and less tolerable with every interaction. A few minutes later her office phone rang. It was Doug Stanton. "Maggie, I just heard we've been rescheduled. Are you going to be ready?"

"Hi, Doug. I've already checked with my team. I think we can have the business review together. What concerns me is being able to lay out the route forward for product development."

"Maggie, I don't understand. You have a very clear route forward using TOC."

"Yes, but we're still missing some of the pieces. I'm used to much faster results in Manufacturing, and I just can't put my finger on what's holding us back."

"Maggie, look at how much more engaged, productive, even creative your team has become with the simple change of focusing in on a few important projects. You just do your best on Thursday, and I'm sure everything will be okay. You know you have my support."

It wasn't Doug that she was worried about.

Storm clouds

Maggie needed to work late again that evening, which was unfortunately becoming a pattern. The business review for the board was ready, but she was still determined to figure out why it was taking so long to get new products moving again. She called home to let Jeff know that she would be another half hour or so, but that she'd pick up a large rotisserie chicken and all the fixings on the way home.

"Well, don't get too much. Sophia came home feeling kind of punky. Feels to me like she's running a fever."

That's it, Maggie thought. *Randy can push me all he wants, but I'm not going to ignore my kids.* "I'll leave right now," she told Jeff. When Maggie got home, she set the food on the kitchen island and went straight upstairs to check on Sophia.

"How are you feeling, sweetie?"

"Not so good, Mom. I'm all hot and my stomach aches."

"What'd you have for lunch?"

"Nothing."

"I hope you're not trying to stay trim by skipping meals," Maggie joked. Sophia was already on the lean side, but girls this age could have a distorted body image.

"No Mom, I never skip lunch. I just didn't feel right."

"Can I make you a little buttered rice? You know with hot-dogs cut up in it the way you like it."

"Yeah, I think that would be okay. Thanks, Mom"

Sophia ate about half of the small plate that Maggie brought her and sipped some juice. "Sorry, Mom, I'm just not that hungry."

"Alright, honey, I'm glad you tried. I'll be back in a bit to check on you. Let me know if you need anything, okay?"

Later that evening after the rest of the family had eaten, Maggie and Jeff both popped in to check on Sophia again. They found Sweetie laying alertly at the foot of the bed as Sophia slept. She still felt a little hot, even after the aspirin, but at least she was resting.

Whirlwind

Maggie awoke with a start, noticing the time on the clock—1:30. She looked to her side to see Sweetie, who was staring intently at her and whimpering softly. Sweetie usually didn't bother them in the middle of the night. "What's the matter, girl, do you need to go outside?" When that didn't get the usual response, Maggie was concerned. She got out of bed and followed as the anxious dog ran into the hall, leading her into Sophia's room. When she stepped in, she found her daughter doubled up in pain on the floor. Maggie felt Sophia's head and she was burning up. This was more than a stomach bug. Maggie pulled on a sweat suit, woke Jeff and had him help her get Sophia to the car so that she could get her to the emergency room while he stayed with the twins.

After a quick, but worrisome ride, Maggie helped Sophia into the emergency room, where she stumbled, threw up and passed out in her mother's arms. As Maggie struggled to hold her up, an orderly immediately sprang into action and placed a wheelchair behind her. A janitor with a mop wasn't far behind. Maggie's apology for the mess was dismissed by his kind, grandfatherly smile. "Now don't you give it a second thought. That girl of yours needs all your attention right now."

From there it all became a blur. They wheeled Sophia into a treatment area where one of the young doctors quickly started the assessment. In just a few minutes, a nurse had her cleaned up and hooked up to an IV. The doctor explained that it could be any number of things, but they suspected it was her appendix or her gall bladder. Someone would be there in a few minutes to perform an ultrasound. While they were waiting, one of the admitting staff quickly popped into the room with a clipboard and said, "Mrs. Edwards, I'm terribly sorry to bother you, but if I could just get a little bit more information on Sophia?" It was almost like the administrator knew that providing Sophia's medical history would help keep her focused while they waited.

A few minutes later, the ultrasound tech arrived with a portable unit and efficiently began the evaluation. "It's hard to see, but that's her appendix right there," he said. "Let me get the doctor back in here for a diagnosis." After the doctor confirmed that her appendix was the problem, she said they wanted to get Sophia into surgery as soon as possible before it ruptured and caused any other complications. In the meantime, they needed a set of x-rays to confirm the diagnosis. They also gave Sophia something to ease the pain and help her to sleep.

With Sophia resting, Maggie stepped into the cell phone area and dialed Jeff. "I was getting worried that I hadn't heard from you," he sighed, obviously relieved that she'd finally called. "How's our girl?" Maggie told him everything that had transpired. They decided that he should drop the twins off with his sister and get to the hospital as soon as he could—before Sophia had to go into surgery, he hoped. "How are you holding up, Maggs?" The funny thing was Maggie would have expected to have felt more frazzled, but the coordination and competence of the staff put her more at ease than she would have expected.

Recovery

Jeff was pacing around the waiting room when the surgeon came in. "Mr. and Mrs. Edwards, your beautiful little girl came through the surgery like a trooper. Sophia's in recovery right now and she's asking for you. As soon as you're gowned, you can go in and see her."

"Gowned?" Jeff asked.

"Don't worry. It's just a precaution. She did well, but she's not out of the woods just yet. We got it early, but her appendix did actually rupture. So, with the fever, we need to keep her for a few days to make sure there's no infection. We don't want any post-surgical complications."

They were both sitting by Sophia's hospital bed when she awoke and managed a little smile. "Mom, Dad. What happened?" Her voice was barely more than a whisper. "I'm really sore. The last thing I remember is getting out of the car."

"Well, you had your appendix removed."

"Will I have a scar?"

"No, honey," laughed Maggie as a tear ran down her cheek—she was so happy that Sophia was well enough to worry about normal pre-adolescent issues. "At least not a big one. They just made a few small incisions."

"You'll barely be able to see them," added Jeff, as he kissed her forehead. "You gave us quite a scare, little one."

Jeff stayed with Sophia while Maggie headed home to get cleaned up and collect the twins. Sleep would have to wait. As Maggie drove, it occurred to her how lucky they had been. They had one of the best hospitals in the country only minutes from their house, and they had performed incredibly in taking care of Sophia. The coordination in the ER had been synchronized almost like a ballet, with every group—doctors, nurses, imaging, even administration and janitorial—playing their part without stepping on the others. She wondered what her organization could learn from them. No, now wasn't the time for that. DFT and Randy Barrister would have to stay on the backburner for a little while longer.

Taking care of business

The following day, Maggie was able to return to DFT. She had resisted, but Jeff reassured her. "Maggie, you have an important meeting coming up this week. The school can find a substitute for me. That's not something you can do."

"But Jeff..."

"But nothing," he said gently pressing two fingers to her lips. "I don't want you to give it another thought. You can pick up the twins in the afternoon and bring them to see Sophia. That'll give you until 3:00 to take care of business. Don't worry. We'll make it all work somehow."

When Maggie walked into the office that morning, her phone rang almost immediately. It was Doug Stanton, calling from out of town because he had heard about Sophia's emergency surgery. After hearing that she was okay and that Jeff would be handling the daytime visits until she was released in a day or two, he asked Maggie if this was going to have an impact on the upcoming meeting.

"No need to worry. I'll figure out a way to be there," she said. She just wished she were as sure about what she was going to say.

"Well, don't spend all day there. You need to take care of yourself, too." That sounded good, but Maggie knew it wasn't very likely as her day quickly evaporated in reviewing the financials and other operating information for the upcoming meeting.

Falling into place

Maggie decided that she'd spend that night in Sophia's hospital room. Who knew how much longer she'd still be a little girl and want her around? Even though her big meeting was the next day, Jeff knew that there was no use arguing. Maggie had made up her mind, reasoning that the meeting wasn't until the end of the day, so she'd have plenty of time to shower and get ready, with enough time left to spend a few hours at the office. She'd be exhausted, but she'd make it work.

There was a DVD player in Sophia's room, so Maggie stopped by the video store on her way. The hospital brought in an extra meal, and they spent the rest of the night on the bed watching one of Sophia's favorite movies until she fell asleep around 9:00. The surgery, even though done laparoscopically, had taken a lot out of her.

As Maggie stroked Sophia's hair, she wondered how the next day would go. Maybe she'd tell Randy Barrister what she really thought of his management capabilities, or lack thereof. *He makes a few bad acquisitions to feed his ego and then tries to make everyone else pay for it. Would it be so bad if I got fired, anyway? I could use the time off any way to spend more time with the kids*, she thought to herself before she drifted to sleep around 1:00.

Maggie awoke with a start at 5:00 a.m., hearing a monitor buzzing in the room across the hall. She realized that someone had placed a blanket over her during the night. The people here were terrific. Within seconds, a team of nurses was hurrying into the other room and another wasn't far behind with an emergency cart. As she watched the emergency play out, it reminded her of the great coordination she had seen in the ER—all the different groups working together so effectively.

"Could that be it?" she whispered to herself. Was that what had been missing in her approach to innovation at DFT? Even the TOC improvements they made in manufacturing had not been in a vacuum. There had often been some coordination required from other groups like purchasing or customer service. But Maggie had been treating R&D and marketing like they were a separate organization, almost as if subordination had meant isolating the constraint from the rest of the organization. Was that part of why she wasn't getting more traction?

She knew she wouldn't be able to go back to sleep, so she flipped open her laptop and searched for something that might help her make some sense out of this.

Last chance workout

True to his word, Jeff was on time—even early—to replace Maggie. On her way home for a quick shower, she called Nancy from the car to let her know that she would be in by nine and wanted her to call the entire team together. "You mean Manny and Danielle's teams?"

"No, all of my direct reports. Yes, even HR and Finance. Oh, and what am I thinking? I'll also need Dean and the rest of the Energy-Saver team in there, too."

Her first order of business at the meeting was to thank everyone for the gifts they had sent Sophia. "I want to let you all know how much all the cards and gifts mean to us. It really cheered up her room and is helping keep up her spirits." After updating everyone on Sophia's progress, she quickly switched gears to the task at hand. "Sorry for the short notice, but we've got a lot to get done this morning."

The group looked around the room, waiting to hear why they had been brought together. "Dean, why don't you take us through your critical chain plan? The rest of you, your mission is simple. I want you to look for any way that you and your team can help the design engineering group shorten or eliminate steps anywhere on the critical chain." Maggie only hoped her inspiration hadn't been a misguided idea born out of sleep deprivation or, worse yet, desperation.

The crucible

Maggie's thoughts were spinning as she paced, waiting to be called into the corporate review session. She realized that she probably wouldn't even be standing if it hadn't been for caffeine and adrenaline, but they were also making her jittery. She took a few deep, calming breaths; she knew she had to focus before facing Randy and the executive board.

After she had completed the financial review, which received mostly positive feedback, Randy moved them to new products. A smug look on his face, he said, "So, Maggie, can you tell us what you've done to get new products at DFT restarted and when we can expect to see some results?"

Well, thought Maggie, *here goes nothing.*

"Randy, let me start by saying that we have made significant changes in how we select, manage and prioritize projects. And we are seeing the results from those changes. In fact, for the first time that anyone can remember, every single project is tracking against plan to finish on-time and on-budget. I'd put those results up against any in the company. But, even so, I agree with you. That's not good enough." She paused, making it clear that she was serious.

"To help everyone understand the approach we're taking, there are a few foundational issues to cover, if I may." Randy begrudgingly nodded his consent.

"When I was in manufacturing, how would you have reacted if I'd come in here and told you that we just needed to make a good product? That we didn't need to worry about delivering more volume each and every year? That we didn't need to continually reduce lead times or reduce operating costs and working capital or inventory investments?"

"I'd probably ask what decade you were stuck in and suggested that a career in government was more your speed," Doug laughed.

"Exactly. No manufacturer in this day and age would survive much beyond their start-up capital without continuous improvement. Even the smallest companies benefit from Lean and Six-Sigma implementations. And all of you know how much extra leverage Barrister Industries gets in the plants where we've used TOC to focus those efforts."

"Granted, but what does that have to do with new products?" Randy asked, unsure where she was steering the conversation.

"Nothing, and that's the problem." She paused, letting that elicit the interest she wanted around the table. "We don't have a systematic approach to extracting more and more from our innovation investment each and every year. We depend on our innovation process to deliver more cash flow in the future. But we've exempted it from the continuous improvement approach that drives success in our manufacturing processes." Obviously skeptical, Randy raised an eyebrow, looking like he might stop this foolishness at any moment.

Choosing to disregard Randy's displeasure, she ticked off the missing elements. "Who wouldn't like to see innovation projects take less and less time each year?"

"I'd just like to see one finish on time to begin with," cracked Tom, the head of corporate marketing, prompting others to nod in agreement.

"How about reductions in development costs?"

Lara said, "Well, as CFO, I have to object to your leading the witness." Her comment elicited chuckles from around the room. "And we're definitively not seeing R&D cost go down. Each division is constantly trying to sneak them up a little each year even as a percentage of sales." Randy still wasn't smiling, but as Lara spoke, Maggie could see the rest of the executive team was gradually warming up to her.

Maggie continued, "And how about cash flow from new products?" The nods were more vigorous now, but his arms crossed defensively, Randy still wasn't coming around.

"Well, in the past few months. I've learned that we just attack these problems one at a time and sporadically—not in any concentrated and repeatable manner."

"I think I see where you're headed," said Tom.

"What I'm saying is that we need to apply the same continuous improvement framework that serves us so well in manufacturing to our innovation process."

"Yes, but isn't innovation more art than science? You can't systematize art," countered Tom. Doug just smiled. He'd obviously heard this argument before.

"I used to think the same thing," said Maggie, "but I realized that was just giving in. Why are some projects successful, while others are complete flops? We owe it to ourselves to find out and use that information to our advantage."

At the whiteboard, Maggie began asking a series of questions. There were some strong egos in the room so she had to help them discover it for themselves.

Her first question was, "Would you agree that the only reason to improve our new product innovation process has to be to help us make more money now and in the future?"

"Well, wouldn't it also improve our brand image as a forward thinking innovator?" asked Tom.

"Absolutely, but ultimately what do you want to get out of a better brand image?"

"More money in the future," he conceded, while others agreed. They were starting to get it.

"Great," she said. "Now that we have a goal, we need a framework for making improvements vs. that goal."

"You mean like the TOC approach we've used in Manufacturing."

"That's right," said Maggie.

"But innovation is such a complex process."

"You're right. Innovation is complex, and it always will be." She paused again, hoping to focus their attention. "But improving it doesn't have to be. TOC was developed for exactly that reason: to simplify the improvement of complex systems. We only need to find the bottleneck: the leverage point where small changes deliver big results. That's the beauty of the approach." She spent a few minutes explaining in more detail, knowing some members of the group weren't very familiar with TOC.

"This is making a lot of sense." A nod of heads indicated that several others agreed with Lara's comment.

With the foundation in place, Maggie continued to outline the bottleneck they found at DFT, the policies and other constraints that were in the way, and the steps they had taken to address it.

"So, you've actually reduced the number of projects," acknowledged Lara again. "Bravo. With too many things going on, it always feels like no one is accountable. I'd much rather see our investment focused on a handful of things where we can really make a big impact, rather than frittered away on lots of little things which never seem to get finished anyway. "

"There's more to it than that, though." Maggie went on to show them that people were four to five times more effective when they didn't multi-task. She also pointed out how sequential pipelining of projects could deliver faster cash flow. Reinforcing her position, she shared how the Electron group had used it for a completely different bottleneck and had several promising new opportunities in the funnel.

"Impressive," agreed Lara.

Up to this point, the discussion was going better than Maggie had dreamed it would. But not everyone was convinced yet.

Back to reality

That was when Randy jumped back in, "Maggie, this all sounds good in theory, but why aren't we seeing more results yet?"

It must have been exhaustion that caused Maggie to sigh with resignation. "Randy, that's my fault." Doug stiffened and the rest of the group went silent. They all knew Randy wouldn't take on someone showing strength, but show weakness, and he could be quite a bully.

But Maggie didn't give him the chance. "With only a few months to figure out this new product thing, I followed my manufacturing experience too closely. I was right about using TOC, because we're actually seeing projects running according to plan for the first time that anyone can remember. However, I was still missing an important piece of the puzzle. Let me tell you about it and what our plans are to fix that."

Maggie paused and looked around the room. "The third step of the TOC focusing process is to subordinate to the pace of the bottleneck. I had interpreted that as the rest of the organization staying out of the way of the bottleneck and timing its action to always keep the bottleneck working."

"That sounds reasonable."

"Yes, it does, but it's so much more than that. Does anyone remember what Peter Drucker said about the role of marketing and innovation?"

A few regarded her skeptically, but Tom finally spoke up. "Wasn't it something about creating customers?"

"That's right, "Maggie was impressed. As she wrote it on the board, she said, "Drucker said *'Because the purpose of business is to create customers,'* we've called that make more money in the future, *'the business enterprise has two—and only two—basic functions: marketing and innovation. Marketing and innovation produce results; all the rest are costs.'*"

"Whoa there," objected Brian, the head of corporate HR. "He may have been considered some kind of a management guru, but I'm not sure I agree at all. We need everyone across the organization to be engaged."

"Yes, but engaged in what?" This time, the question came from Doug. "Isn't it wasted effort if it doesn't somehow enable or improve the customers' experience or satisfaction? Shouldn't every person in this company somehow be supporting one of these two basic functions?" lightly tapping the table for emphasis.

"Well, Drucker's statement seems to be trying to exclude or classify other groups as non-value added," Brian objected.

"Yes, but what if he meant that to add value, every job should either result in or support the creation of a customer?" asked Doug.

Maggie had an idea. "You're right, Brian. Drucker's words could be taken out of context. Let me try to capture this differently in a way that more people might accept." Walking back to the board, she wrote:

Innovation:
The organization-wide process for finding and profitably serving unmet customer and market needs

"Drucker wasn't talking about the marketing and innovation groups. He was talking about the activity of marketing and innovation. Everybody in the company needs to be part of either the marketing or the innovation activity," Maggie stated.

"Now, that I like," said Brian. "Every part of the organization. That's really important."

"And profitably serving is really important, too," added Lara. "Sometimes we wait too long to consider that and end up with new products that we can't afford to sell."

"Let me ask you something though, Brian," said Doug. "Could you see others who don't work directly with customers, like production workers, relating to this definition?"

"Definitely. Aside from the obvious critical role they play in making new products, any new ideas they could contribute on how to simplify production and improve throughput would be part of profitably serving. I think the same thing would apply to almost any functional group, whether that be Purchasing, HR, Quality Assurance, you name it."

"Maggie, what does this have to do with my question about not getting results faster?" asked Randy, impatiently tapping his foot.

"Actually, it contains the missing piece that's held us back from making even faster progress at DFT," said Maggie.

"I don't see..."

"Randy, I told you it was my fault. I was so focused on getting the Marketing and Product Development groups moving again, that I underestimated the role the rest of the organization must play. I was making sure the rest of the organization was subordinated to Design Engineering's pace and was ready to run with projects as soon as they finished tasks. But to really be successful, I believe the Subordinate step has to be more than that. The rest of the organization has to be actively involved in helping the bottleneck by either helping them do their work or, even better, by figuring out how to eliminate work for them.[25]

"Maggie, I don't see how this is going to make a difference," said Randy, obviously frustrated with her.

"Well, Randy, I'm sure you are aware that we've been struggling with a project timeline that we could not shorten up any further, and as a result, we were at risk of losing some of the business at Midland—one of our largest customers."

"Painfully aware," he grumbled.

"Well, I just came out of a meeting where we were able to eliminate several steps that had been holding up the project and preventing us from meeting Midland's target. By getting the rest of the organization involved, our purchasing group identified two off-the-shelf solutions that the engineers hadn't found. So we've eliminated two design tasks and replaced them with a systems integration task that doesn't involve the bottleneck."

Randy didn't seem to know how to respond, so he just stammered, "Nice job—by your team, I mean..."

Doug stood and said, "Indeed, a very nice job by you, as well, Maggie." Turning to Randy and the rest of the group, he proposed, "I think what Maggie's laid out here makes a lot of sense. We owe it to ourselves to let her see it through. Admittedly, it's early yet, but if things work out like I expect at DFT, we're going to want to talk about rolling this out in the other divisions, too. Let's not hold Maggie up any further. She's got a daughter getting out of the hospital soon and lots of work ahead of her implementing this approach. Let's meet again for another review at the one-year mark. How does that sound?" With solid agreement around the table, he asked, "Randy, I assume that that you are okay with that approach?"

"Well, I'd like to hear more about how you're going to continue the improvements from here, but yes, I guess so.

"Great," said Doug. "I know Maggie is plenty busy implementing many of the actions from the Exploit and Subordinate steps, but I'll follow up with her later on what she'll be doing for Elevate."

And with that, Maggie had the time she would need.

 **Step 4 – Elevate**

Breathing room

"See, I told you there was nothing to worry about," Jeff said as they drove to the hospital that night to pick up Sophia. "You had it in the bag all along."

"Well, I'm glad you thought so," sighed Maggie. "But honestly, it wasn't until this morning that I had anything to tell them."

"Well, inspiration strikes in the strangest places." They both smiled, remembering the story she had told him about the emergency room visit that had reminded her that even in the planning stages, improving innovation took more than the marketing and R&D functions. "But it also takes a special kind of courage to grab it and make something happen. I just hope the twins don't have to have their tonsils out the next time Randy puts you in the hot seat." he teased.

"That's not funny," she said, trying to stifle what she knew was an inappropriate laugh. "But at least I've got some time now." Of course, in the back of her mind, she knew that the rest of the year would go quickly if she didn't stay focused.

Not there yet

One month later - Maggie smiled at the pride and owner-ship her leadership team had as they celebrated the news. DFT had just completed the first prototype for the Energy-Saver system, and Midland had been so pleased with the initial per-formance reports that they were already talking about pushing up their replacement plans. Maggie reminded the team there were still a few hurdles to get over first. But, on the inside, she was celebrating, too.

"Alright, thanks again for that news, Danielle. I have some other news that you should all be aware of, too. Roger has an-nounced his retirement and is going to be finishing up at the end of the month." Maggie forced herself not to smile as she said it, but she could tell by the looks around the table how re-lieved everyone felt. "So, unless anyone else has any pressing news, we're all here today to talk about what our next steps are going to be to continue improving our new product perfor-mance."

Maggie reviewed the progress they had made so far and where they were in the process. "And you'll notice that this time, I've included the entire team, not just R&D and Market-ing. I know some of you may still question how you can help, but remember innovation is a cross-functional sport."

"Don't worry, we're all onboard after seeing how we were able to help with Midland," said Harvey, the HR lead.

"Great. Well, I think we may be at the point where we need to start thinking about elevating our capacity, so I'd like to hear your thoughts on how we can add capacity to our Design Engineering function," said Maggie. "If you recall, during the Exploit step, we focused on policy changes and the invest-ments required were minimal—next to nothing."

"High-leverage changes as you put it," said Jean, the head of finance. "My favorite kind."

"That's right," continued Maggie. "What's different in this step is that we will have to make larger investments to get that capacity."

"Do you mean headcount and equipment?" asked Jean.

"Possibly, but even here, I'd still like to start with changes that don't add significantly to our ongoing fixed costs. I'm not saying we won't eventually get to that point, but we'll start by looking for the highest leverage actions to elevate the capacity of the resources we already have."

The group brainstormed and quickly settled on a list of alternatives worthy of evaluation:

- Training
- Design of Experiments, Modeling and Simulation Tools
- Selection, Coaching, and Development
- Software Systems
- Outside Resources/Open Innovation

"Okay, let's start with training," said Maggie.

"I'm confused," said Jean, the head of finance. "Didn't we have the Professor in for some training as part of Exploit?"

"Fair point, Jean," replied Manny. "I think it really depends on where your bottleneck is, but for Design Engineering, we identified that in order to exploit all of our capacity, we needed a different project planning process. So training was required to implement that new process. In the elevate step, I think we're talking about skills training that adds capacity. Does that sound right?"

"I'm not so sure about that," said Danielle. "If Exploit is about working smarter and Elevate is about adding capacity, then I think most training is Exploit."

"That's a good observation." Maggie was continually impressed by Danielle's progress. "Since TOC was primarily developed for manufacturing, we're going to have to decide how to apply it as we go along. But the most important thing is to be practical about it. Let me suggest a definition that may be more appropriate to knowledge work, where I think it's sometimes too easy to lump everything into working smarter, retrospectively."

Maggie stepped over to the whiteboard as she continued, "If Exploit is about working smarter to get more from the capacity that's already available, then Elevate is about adding or creating new capacity."

Everyone was quiet for a few seconds while considering what she had written.

"I think that works," said Danielle.

"Okay, so Manny, how do your Statistical Design of Experiments training ideas fit into the picture?"

"Not to muddy the waters, but I think it would be Subordinate," replied Manny. "Because we're directing non-constraints to work smarter by eliminating work for the bottleneck with an optimized experimental plan. With a small investment in software, a little training and a policy change, we could cut as much as 10% of the work in Design Engineering by eliminating a design iteration or two. But should we really get so hung up in what we call it?"

"It only matters what we call it to the extent that it helps prioritize," said Maggie, "Since we generally give earlier focusing steps priority. But whatever we call it, let's get it done this quarter."

They discussed some other skills training, but none seemed to offer a throughput increase in DFT's situation, so Maggie moved them along to software.

"Well, under the category of Exploit, I've seen the design engineers calling around to get the project information they need more quickly," remarked Danielle. "I thought it could be a timesaver if we could begin using some of the low-cost web-based collaboration tools that are out there to create virtual team rooms where the project manager could keep all of the project files. It's just one less step they'd have to take to find the info they need."

"That's definitely working smarter," observed Manny. "Wow, I feel bad that I missed that one earlier. Sounds like an unnecessary headache"

"Maybe so," said Maggie, "But it certainly wasn't the biggest problem we were facing at the time, and the key is to find the change which gives us the most leverage at that point in the improvement process—the most bang for our buck, as they say. What do you think it would take to implement virtual team rooms, Manny?"

"Well, I could use some help from the IT group here," he said.

"The easiest approach may be the team room functionality already built into our email software," said Jean, who was responsible for IT. "But I'll have someone start looking into the options right away."

"Under the category of working smarter, what about project management software?" asked Manny.

"We already have MS Project on all the engineers' computers, don't we?" asked Jim, the supply chain VP.

"No. I'm talking about critical chain software that lets you schedule and manage resources across projects."

"How big an issue is that, Manny?" asked Maggie.

"Honestly, it's more helpful to do it manually right now while we are learning, but eventually it would allow us to handle more projects simultaneously."

"So maybe it's something to start considering for next year's budget," said Jean.

"That sounds about right."

"Now, what about modeling and simulation?" asked Maggie.

"Well, I think modeling and simulation is definitely an elevation of our capacity," Manny replied. "Virtual testing is working smarter, and will require training, but this is more than just adding a software tool. We'd be adding capacity by building a capability that we don't currently have. It will take a little longer to create the initial designs, but we'll get to working prototypes faster and that will add about 20%-30% more to our engineering capacity."

"Sounds promising, Manny," said Maggie. "Can you get an estimate for what it's going to cost and how you would implement it with minimal downtime? I'd like to have you review it at our next meeting."

"Yes, that should be enough time. Though I could use some help from Jim's purchasing folks to spec out and negotiate the workstations."

"Glad to help."

"Let's move on to the people side of the equation," continued Maggie, asking Harvey to explain further.

"Studies show that people are far more effective if they are in a job that allows them to use their strengths.[26] We're already working to make sure people have the right skills training, but

I'm talking about something different here," Harvey paused. "I'm talking about people having the right competencies for the job they are in. Are all of the people in Design Engineering A-level talent for the jobs they are in or are they capable of becoming top-level talent and making progress in that direction?[27] If they aren't, we can elevate our capacity by moving that kind of talent into those roles. What do you think, Manny? Do you have the right people?" "

Manny looked a little hesitant, so Maggie stepped in to reassure him. "Let me remind everyone that as part of the DFT leadership team, what's said here regarding personnel issues must remain confidential."

"Well, in that case," began Manny, "I do have at least one person in the group that I think might be in the wrong job. Hank Deveraux is a hard worker and a solid employee, but he doesn't quite have the eye for design required to be A-level talent. Sometimes, his work ethic helps makes up for it, but still…Well, let's just says he doesn't have the same capacity as some others.

"Is there anyone else in the company with the right skill set?" she asked.

"No, not in any of my groups."

"Harvey, any ideas?"

Before he could answer, Jim offered. "Maggie, I may have someone in the Manufacturing Engineering group that would be a perfect fit. Jack Dugan has always had quite a talent for this type of work. Honestly, I think he's getting a bit bored with manufacturing support. Plus, he has a simulation background."

"Sounds like swapping positions could raise the capacity in Design Engineering and have the added bonus of putting both employees in a better position to utilize their strengths," said Harvey. "Although we'll need to be careful so that Hank views this as an opportunity, too."

"I love win-win. Harvey, I'd like you to work with Manny and Jim, get some developmental discussions going with each of them and come up with a recommendation before our next meeting.

"The last item was Open Innovation," said Maggie. "Somehow, I have a feeling that is going to be a little more complicated."

Not ready for open innovation?

When they approached the topic of open innovation, Maggie admitted she just wasn't comfortable going outside yet. "In fact," she said with some hesitation, "while it makes me feel like the Grinch, I think we should put this one on hold—not for too long, but at least for now."

Seeing a look of frustration on Danielle's face, Maggie quickly explained. "I think OI has great potential as a way to elevate our innovation capacity, and I know there's lots of excitement about working with outside groups, universities, and even government agencies to find a solution to the problem we ran into on the Ultra-Compact System. But, before we do that, I want us to be ready and to really have our own act together. It wouldn't be fair to the partner nor to ourselves. We've made a lot of progress, but we really need to deliver on the first handful of projects before we add the complexity of a partner."

"Well, I guess that's fair enough," said Danielle, somewhat appeased.

"Is it really so complex?" asked Harvey.

"I've never shared this before," admitted Manny, "but in my last company, we had an experience along these same lines. We put together an outside consortium to develop a new metering technology and within six months of getting the deal done, we were on the verge of a break-up, which happens to about 70% of external alliances. We bought in some experts, but it was too late. Unfortunately, the alliance team thought we were headed in one direction, while factions within each company had their own views—none of which were even close to being aligned."

"That doesn't sound very encouraging," said Harvey.

"It was a bad situation, but it doesn't have to be like that. Corporate alliances can be like Vegas marriages if you aren't careful, but there are some really good frameworks out there for building lasting partnerships.[928] Maggie's right, we need to be sober about it and make sure our house is in order first."

The Story

Part IV – Turning the Flywheel

Step 5 – Start Again

Building momentum

Three months later - "Danielle is on a conference call with Midland, so she's going to be a few minutes late," Maggie informed her leadership team. "Keep your fingers crossed that she gets good news. Meanwhile, let's get the meeting started."

"Maggie, I'd like to start by apologizing because some of the work we did in Elevate might have been wasted time," said Manny.

Thinking it had been anything but wasted time, Maggie was confused. "Manny, what in the world are you talking about?"

"Well, I've been going over the data, and it turns out that we were getting pretty close to moving the bottleneck before we even made the changes last time. With the changes we've started making to add modeling and simulation and the groups growing skill using it, I estimate that the bottleneck will move to the drafting group sometime in the next quarter. They'll no longer be able to keep up with the design engineers."

"Moving the bottleneck?" Jean's confusion showed on her face.

"Yes, sometimes it's called breaking the constraint. But the point is that eventually you eliminate one bottleneck and move on to the next."

"So what's the matter?" asked Jim, "Drafting is an easily scalable resource."

"Nothing's the matter. I'm confident that we can continue to increase our throughput and even reduce cycle time further, but it turns out that incorporating DOE alone would have resulted in outstripping drafting. The other steps weren't necessary."

"Well, that's the kind of wasted effort I like to hear about," laughed Maggie. "Although, I'm sure there's a lesson in there somewhere."

"I think so," mused Manny. "Before proceeding to the next step, we always need to check how close we are to eliminating the bottleneck. But you know, something still doesn't seem right about letting the bottleneck move to drafting anyway."

"What do you mean?" asked Maggie.

"Well, from a constraints management perspective we may want to decide to leave the bottleneck in Design Engineering."

"Wait a minute," objected Harvey, "Why would we want to do that? Wouldn't it stop future improvement?"

"No, quite the opposite," replied Manny. "Let me explain.

If having the constraint in Design Engineering provides the best drum beat for coordinating the rest of the organization, we might want to keep it right there. We'd still focus on improving the constraint, but occasionally, we would also need to upgrade the non-constraints so that they keep up and have enough protective capacity.

"Okay, so that's why you called it constraint management," said Harvey, "We're deciding the best place to manage the constraint and then not only using it to set the pace for the innovation process but also for improvement."

"Exactly," said Manny. "In fact, we should talk about whether it makes sense to manage Design Engineering as our constraint, which I think it might. If so, then we'll need to explore some ways to work smarter or add protective capacity in drafting so we can rebalance the workflow."

Unexpected bonus

Maggie had been waiting anxiously for Danielle to join them. Everything had gone well throughout the prototyping evaluation, and Midland's first system was installed and operating almost two weeks earlier than promised. Today, they were supposed to hear what kind of energy saving results Midland had obtained. When Danielle finally finished the conference call and joined the meeting, she furrowed her brow and quietly sat at the other end of the table from Maggie. It didn't look promising, so Maggie stopped the discussion. "Sorry to interrupt, but the suspense has been killing us. Come on, Danielle, let's just get it over with."

Danielle stood up and winced, as everyone seemed to steel themselves for the bad news, then a playful grin erupted on her face as she threw an email copy of the purchase order on the table, punched a fist in the air and shouted, "We nailed it!"

Not only had the results been better than promised, but Midland had placed orders to replace all of their systems over the next two years. With similar results expected from several other big accounts, the plant would be sold out by year-end.

Maggie could have kicked Danielle for her little stunt, had she not been so busy celebrating herself. "Well, I suppose after working so hard on this bottleneck, we deserve a little rest," she said playfully.

"What, are you kidding? No way, we've got momentum—let's keep going," the group replied, almost in unison.

Now that's what I wanted to hear, thought Maggie.

She marveled at the turn of events, remembering how she'd always worked hard in manufacturing to prevent production from becoming the bottleneck. Then with the move to DFT, her focus had changed to eliminating the market constraint. Now things were coming full circle, and it wouldn't be long before they'd be focusing on elevating manufacturing capacity again.

Interesting, she thought, *the whole thing works together almost like a flywheel.*

Back on the fault line

One year later - Maggie was concerned as she entered the country club. Doug had called the lunch meeting just this morning, and now she found him sitting here with J. Randolph Barrister III. It had been over six months since her one-year review session with the executive board, and she was happy that meeting had gone even better than the first. Luckily, she hadn't seen much of Randy since. Thinking her luck must have run out, she figured she'd done something else to upset him, though she couldn't imagine what it could be. In her opinion, things at DFT had been going well and the results certainly confirmed that, but that never seemed to be good enough for Randy.

"Maggie, I'm glad you could join us," said Doug. He always had her back, but his face wasn't giving anything away today. She noticed a bottle of her favorite varietal was already chilling beside the table. *Wine at lunch?* she wondered.

"Don't worry, Maggie, I'm not here to pick a fight," Randy said, obviously sensing her discomfort. "Besides, you always seem to be able to hold your own. In fact, that's part of what I've come to respect about you, even if I haven't always shown it."

"Um...well, thank you," stammered Maggie, a little dismayed.

After hearing their server recite the specials, they all ordered and then made small talk over a glass of wine. To Maggie's relief, the meals arrived quickly. Maggie commented on how good her salmon salad was, and then knowing that they wouldn't be interrupted for a while, she broke the ice. "I'm sorry, gentlemen, but I have to ask. To what do I owe the honor today?"

They both looked at their watches, and then Randy handed Doug a 20-dollar bill. Doug shrugged and said, "We had a bet – how long it would take you to ask. I had the under."

Well, she was glad that she'd been able to entertain them, but that didn't answer her question.

Doug must have noticed her look of expectation. "Oh yes, the reason we're here," he laughed. "Well, Maggie, it's no secret that your division has jumped to the front of the pack when it comes to new product results, and it's evident that the help you've given Chuck at Electron has also made quite a difference there, as well."

"I'll admit, too, that while I might have been biased toward growth by acquisition," added Randy, "you've helped me see the critical role that organic growth can also play for Barrister. More importantly, you've shown me how a constraints-based innovation framework simplifies our choices and helps us to maximize results without adding costs."

"Maggie, what we're trying to say is we'd like to ask you to think about changing jobs again."

"But, Doug, Randy," she implored. "I feel like we've only gotten started. There's so much more we can accomplish if we just..."

"Oh, we're counting on that," Randy stated with a measure of amusement. "We're asking you to take over as Chief Operating Officer.

"I didn't realize..."

"You'd still report to me," Doug said, taking over the conversation on cue. "But you'd be responsible for driving organic growth in each of the operating divisions. I'll be shifting my focus to the acquisition and financing side. And, I can tell you right now, one of my objectives will be to find companies that can benefit from the system that you and the Professor have created."

They continued to discuss Doug and Randy's vision for the remainder of their lunch. The possibilities were exhilarating, and Maggie could see that it would be a wonderful opportunity.

Joint decisions

On her way back to the office, Maggie called Jeff and got his voice mail. She left a message asking if he minded dropping the kids off with his sister for the evening. She had something special she wanted to discuss over a nice dinner.

Jeff called her back an hour later, obviously curious about her message. "Maggie, what's this special thing you wanted to discuss? You're not...well, you know."

For a second she was stumped and then realized what he was hinting at. "Pregnant? No, absolutely not. We agreed to stop at three."

"Well, then, why so secretive?"

"It's a new career opportunity that I need to discuss with you, and it'll be easier if we have some peace and quiet."

"Oh, is that all? Look, Maggie. You know I'll support you in whatever you decide."

"I know, but still..."

"Don't worry, Maggs. I'll drop the kids off and see you about six."

A different leverage point

Over a sumptuous dinner of steak grilled to perfection and Jeff's favorite baked sweet potatoes, Maggie explained the offer Doug and Randy had given her at lunch.

"Wow, that doesn't sound like Randy. You've really turned him around, haven't you? So, that's what this is about—taking the COO position. What's your concern, Maggie—the travel?"

"No, most of the divisions are only a short flight—in and out same day. But, the Barrister opportunity isn't what I wanted to talk about, anyway."

"Now, I'm totally confused and after only one glass of wine."

"Two," she teased, "But who's counting?"

"Evidently you are," he laughed. "What's the opportunity?"

"Well, you know that the Professor is retiring at the end of this school year," she explained. "On my way back from lunch, I stopped at his office to discuss the whole situation with him. He holds a lot of clout around the college, so he asked the president to meet with me. We hit it off really well, Jeff. He's liked having the Professor there as a constraints management expert, but he's always dreamed of establishing the college as a center of excellence in the area. With the professor retiring, he had given up on that dream. That's where I come in." She smiled as Jeff listened patiently.

"Of course, I don't have a Ph.D., but it's not an academic institution, so they can be a little more practical. That's one of the things I really like about it. They're more hands on. Leading a department there would give me the opportunity to make so much more of an impact. I could leverage the system of improvement we've created at DFT, not just to other divisions in Barrister, but to other companies locally, and who knows how far when you consider where our students could take it. Of course, it wouldn't be anywhere near as much money as the COO position, but I'd be able to consult, too, so it would work for us financially."

"You sound pretty excited about this, Maggie."

"There are still a number of other folks on the selection committee that I'd have to meet, and it would certainly help if I could get Doug and Randy to fund a department chair. But, yeah, Jeff, I am excited."

Fast forward

As Maggie looked back over the last few years, a lot had happened since that fateful afternoon at Barrister's headquarters. She had saved her job by taking on a new role. In doing so, she'd learned a lot about new product innovation and about convincing skeptics by achieving real results—the only ones that count. As that thought passed her mind, she was reminded of the twins and their struggles with math. *Oh, yes,* she thought, *new approaches can be challenging, but they can also create growth.* She was pleased with the twins' progress in math, and just as pleased that Randy had finally seen the value of organic growth for Barrister Industries.

In spite of the obstacles and challenges they'd faced, it had been a good year. The house renovation was nearly complete, and Maggie was glad to have a job that enabled her to spend more time at home with Jeff and the kids. They were growing up much too fast, and Sophia would be entering high school in the fall. She looked forward to sharing that time with her, shopping for dresses for school dances and, all too soon, exploring colleges together. It looked like she and Jeff were going to be able to give them that top-quality education, after all.

Back at Barrister, Doug and Randy had been shocked when Maggie first shared her alternative vision with them. However, Randy quickly warmed to the idea when he realized that they could still have Maggie's help and that even with a generous consulting retainer, it would actually cost him less than promoting her to COO and hiring a replacement. Over the course of a few months, they had worked out all of the details, and Maggie became the Barrister Industries Chair of Continuous Improvement & Innovation Studies. With a title like that, Doug had teased her that she was going to have to get a bigger business card.

Maggie had also been delighted to see Danielle step into her old position as GM for DFT. She was young, but she had learned a lot in a very short time. Working together with Manny, who had developed a deep competence in innovation management, the strong results they continued to deliver had interested all of the divisions in learning what they could do to keep up.

Over that time, Maggie continued to help roll out the system within all of the Barrister companies—the results made their investment seem insignificant. At the college, Maggie was also thrilled to add a class on constraints based innovation leadership to all of the others that the Professor had long taught. The new program was a resounding success. Word had gotten out about the continued improvements at Barrister and in its first semester, she already had a long waiting list. Maggie smiled when she heard that, delighted at the prospect of attacking another bottleneck.

Epilogue:
Putting the guided innovation system to work in your business

Now that you've read the story of Barrister Industries and learned how a constraints-based approach can simplify innovation improvement and maximize new product growth, perhaps you'd like to explore how you can achieve the same kind of results with your constraints.

While Barrister is a fictional company created to illustrate certain issues, your company is a real one that faces its own unique bottleneck with its own set of issues that constrain further growth. Unfortunately, it's impractical to cover more than a few of the possible bottlenecks and constraints in this book. There simply aren't enough pages for all of scenarios that you could be facing.

That's why we developed the Guided Innovation System— to provide you with the tools to identify and systematically exploit the unique bottleneck constraining your new product growth.

The following section, Part V, provides a brief tutorial on the Theory of Constraints and a 5-step overview of how you can put the Guided Innovation System to work in simplifying the process of improvement and maximizing the results from your new product development process.

Simplifying Your Innovation

Part V

The Theory

Theory of Constraints is an unfortunate choice of terms since TOC is anything but theoretical. In fact, for just this reason, some practitioners refer to it as constraints management.

TOC basics:

TOC is based on the simple principle that the output of any system is constrained by its lowest performing element—also known as the process bottleneck or the system constraint. Since that is what restricts the system's performance, improvement efforts only deliver results when directed at the constraint. That can be by operating it more effectively or by adding capacity. This concept creates tremendous leverage because nearly 100% of the results come from efforts taken to improve or protect the bottleneck.

Traditional improvement processes don't distinguish between constraints and non-constraints and often try to improve the entire process. However, as Goldratt said, "*An hour saved at a non-bottleneck is a mirage,*" since the bottleneck still limits the system's output.[9] The illustration in Figure 1 provides a simplified example.

Figure 1

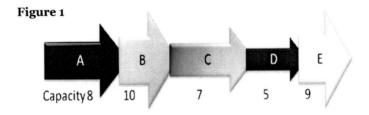

Operating steps A, B, or C at any rate higher than five units will only cause work to pile up in front of step D. Similarly, improving Step E, with a capacity of nine units, will not increase output since D can only operate at five units today. This means that Step D is the constraint or leverage point. Improvements to D will immediately increase output up to a maximum of seven units. Above seven units, Step C becomes the new constraint and the next area of focus.

Establishing the goal:

The improvement cycle starts with establishing the goal for the system that we intend to improve. When a business is the system, the primary goal is usually to make money now and in the future.[9] Some may argue that satisfied customers or secure, engaged employees should be the goal. These may be necessary conditions, but to survive, businesses must continually generate more cash than required to make and sell the goods and services they provide. Moreover, that's the yardstick used to evaluate any improvement: making money now and in the future.

TOC also provides a simple metrics for improvement by introducing the concept of sales throughput—essentially the cash flow from sales less truly variable expenditures, such as raw materials, components, most outside purchased services and commissions. We use cash flow measures—not profit or cost accounting measures that distort the picture with allocations.

The five focusing steps:

Now that we've identified the goal and the metric, how do we drive performance? We want to see real results that increase sales throughput. Businesses are complex systems with inputs of ideas, cash, raw materials, and labor and an output of sales. Even the simplest business ends up having a multitude of levers that affect performance. In a process of ongoing improvement, TOC provides five focusing steps:[16]

1. **Identify** – *Agree on the system's goal, identify the bottleneck, and what is constraining it*
2. **Exploit** – *Decide how to operate the bottleneck in order to use as much of its capacity as possible*
3. **Subordinate** – *Operate the entire process at a rate which optimally exploits the constraint, and use any*

excess non-constraint time to help or eliminate work for the constraint

4. **Elevate** – *Add or create capacity at the bottleneck*
5. **Start Again** - *Don't let inertia become the next constraint*

Before moving from step to step, it is important to assess whether actions taken in the previous step have already eliminated the constraint. Technically speaking, we can't break the constraint without increasing its capacity. However, constraints sometimes hide the bottleneck's real capacity. If the constraint has been broken, continuing through the other steps would be wasted effort. We bypass the next steps and go straight to Step 5 to identify the next constraint and continue the process of ongoing improvement.

In some situations, it makes sense to manage the constraint rather than eliminate it. In this case, it is important to monitor utilization of the non-constrained resources and occasionally raise their capacity, as well. This provides the protective capacity necessary for a steady flow to the bottleneck.

Proven results:

So, where is the evidence that TOC works in practice? Until recently, most of the evidence for TOC has been anecdotal examples from more than 25 years of successful real world implementations in everything from small companies to the operations within larger companies like DuPont, General Motors and AT&T. However, just last year Sanmina-SCI, the $11 billion electronics component and assembly firm, released the first comprehensive study with statistical evidence of TOC's impact.[6] They conducted an internal evaluation comparing traditional Six-Sigma and Lean implementations with TOC used to focus the Lean and Six-Sigma tools. The results were stunning. While all 21 plants saw improved results, the 6 plants using TOC to focus efforts generated 89% of the total improvement. The other 15 plants generated only 11% of the results. By focusing on the constraints, TOC leveraged improvement efforts to produce 15 times greater improvement than traditional methods alone.

The Guided Innovation System

Applying TOC to your new product development

Traditional improvement efforts work on the premise that strengthening each process step makes the whole process stronger. Unfortunately, in most cases, this dilutes your efforts, which is contrary to your goal. TOC takes another approach, recognizing that processes are interdependent chains of activities. Of course, the strength of a chain is not determined by the sum of its parts, but by its weakest link. To get the maximum leverage from your efforts, you must focus on the area where it will have the most impact: on strengthening the weakest link in your innovation chain.

The Guided Innovation System extends these TOC concepts from their manufacturing roots to the area of new product and service innovation. It provides a powerful tool for leveraging your new product efforts to create growth by finding and exploiting your innovation leverage point—the place where small changes will have a big effect. In the following pages, we'll introduce you to five powerful focusing steps for attacking the issues that constrain your organization and wringing significantly more bottom line impact out of your innovation investment.

Five steps that drive innovation improvement

Whether you are in consumer or industrial markets, whether you provide products or services, improvement starts by finding the innovation bottleneck – the step in your innovation process that is holding back faster growth. Then five focusing steps help to eliminate that constraint and allow you to move on to the next, with an increase in innovation throughput as each cycle of improvement is completed. Let's start by taking a closer look at what is required in figure 2.

Figure 2

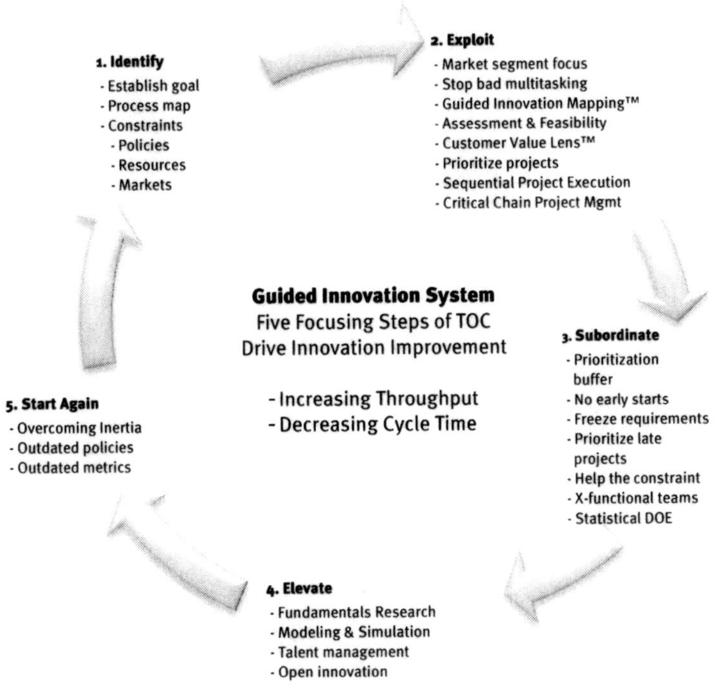

1. Identify
- Establish goal
- Process map
- Constraints
  - Policies
  - Resources
  - Markets

2. Exploit
- Market segment focus
- Stop bad multitasking
- Guided Innovation Mapping™
- Assessment & Feasibility
- Customer Value Lens™
- Prioritize projects
- Sequential Project Execution
- Critical Chain Project Mgmt

Guided Innovation System
Five Focusing Steps of TOC
Drive Innovation Improvement

- Increasing Throughput
- Decreasing Cycle Time

3. Subordinate
- Prioritization buffer
- No early starts
- Freeze requirements
- Prioritize late projects
- Help the constraint
- X-functional teams
- Statistical DOE

5. Start Again
- Overcoming Inertia
- Outdated policies
- Outdated metrics

4. Elevate
- Fundamentals Research
- Modeling & Simulation
- Talent management
- Open innovation

Some prerequisites

Improvement starts with a goal. What are you going to improve? Your current operations are focused on the goal of making money today. So the goal of improving innovation must be to generate more money in the future—more money than your operations would generate without new products and services.

Of course, you also have to know what metrics you will use to measure improvement. While you can measure innovation performance in many ways, throughput and cycle time are key. In the previous chapter, we defined throughput as the cash flow from the sales of new products and services less truly variable expenditures, such as raw materials, components, commissions, and some outside purchased services.[16] Increasing throughput means more money or a higher return from a given investment. Cycle time is the length of time it takes for a project to go from accepted proposal to generating throughput. Decreasing cycle time means that your innovation begins paying off sooner. But, notice that these concepts are connected. Taking on easy projects to reduce cycle time won't help you if it reduces growth in new product sales throughput.

STEP 1. IDENTIFY YOUR CONSTRAINT

What is preventing you from achieving higher innovation performance in terms of throughput and cycle time? When you ask the people in your organization to help identify the constraints, you'll probably find no shortage of issues. But the key is to find the system constraint—the step that is limiting the output of the rest of your system. Before getting started, it may also be helpful to review the following sections on process mapping and understanding constraints.

Process mapping

Having a map of your innovation process is critical because it allows you to understand where the process is breaking down.

- Are you getting too few new ideas from the marketplace?
- Are projects frequently delayed in development?
- Are customers slow to adopt your products?

With a process map, you can ask team members across all functions where they see the symptoms that help identify the bottleneck.

- Where do the biggest delays occur?
- Where does there always seem to be a backlog of work?
- Where are downstream groups constantly idle and waiting?

Mapping your process doesn't need to be complicated. The basic idea is to sketch out how a new idea comes into your company and eventually becomes a product or service reality. Gather a few of your more seasoned commercial and development people—the ones who really know how to get things done. Ask them to diagram how your latest and best-selling products were commercialized, and you'll have a good picture of the current process.

Understanding bottlenecks and constraints

While the terms are often used interchangeably in many situations, including in this book, a bottleneck more often refers to a step that limits a specific process, and constraint refers to a limitation that affects the entire system. Constraints can occur internally or externally. Many companies find that they have a development constraint inside their organization—meaning they have more projects and ideas than they can successfully execute. Others struggle with not having enough ideas or not enough high impact opportunities coming into the organization. Regardless of what constraints your company faces, they would generally fall into one of the following three categories: policy, physical and market.

Policy or management constraints:

Policy or management constraints are the constraints that organizations unknowingly or voluntarily place upon themselves. As the law of unintended consequences predicts, companies can be very creative in adopting policies that inadvertently constrain growth, and policies make up the bulk of constraints affecting organizations. Examples include travel restrictions that reduce customer interaction and multiple levels of review that delay project approvals. Unfortunately, companies can hold policies quite deeply: even if no one can remember the original reasons for their implementation.

The good news is that if you find that your bottleneck results from a policy constraint, rapid improvements are possible. Of course, companies can be blind to their own policy constraints. Sometimes it takes an outside observer or a change in personnel to help challenge the status quo and facilitate the change. In addressing policy constraints, it is important to keep everyone focused on the overall system rather than individual or departmental concerns. Sometimes, it isn't until people see the negative impact on total throughput that they can agree to put departmental considerations aside.

Physical or resource constraints:

A physical or resource constraint simply means that your staff, equipment, or even facilities resources don't appear to have any more capacity. In particular, shared resources are frequently a source of conflict and the bottleneck that delays projects across the entire organization. However, your response should not be a knee jerk reaction to run out and hire additional staff or buy more equipment. When a particular machine is a manufacturing bottleneck, TOC companies don't immediately try to solve it by buying and installing another machine. They defer that investment and first attempt to do more with current resources. Likewise, you should first maximize the output of existing innovation bottleneck resources before considering additions.

Market constraints:

A market constraint simply means that you have more production capacity than the marketplace requires. Product development and marketing are constantly trying to eliminate market constraints by creating new products, finding new applications and markets, and generating sales leads. If you aren't facing a market constraint, then you should focus more of your improvement efforts on your production process since that is where the fastest gains will come.

Of course, markets change and innovators must constantly be aware of developments that could make their current offering obsolete. For example, the fundamental shift to digital media is creating a market constraint for the commercial printing industry, while also creating numerous new markets for search-based advertising and digital printing.

Market constraints can also be policy constraints—the result of a company's choices. For example, what about products described as being ahead of their time? Doesn't launching a product that is ahead of its time mean that internal policies and choices have failed to reflect the market reality and instead have let inertia carry projects along, while continuing to allocate scarce R&D resources to work on a product that wasn't needed?

STEP 2. EXPLOIT THE CONSTRAINT'S AVAILABLE CAPACITY

Now that you've identified your constraint, what can you do to fully exploit its capabilities?

Create focus with uniform process & assessment

Without a concentrated effort to weed out underperforming or lower promise opportunities, 80% of new product results will tend to come from only 20% of your projects.[22] Furthermore, the bottom 50% of projects will only contribute 5% of your results. This means that you gain tremendous leverage with even a modest shift in focus to better opportunities.

To illustrate this effect, one study showed the dramatic differences between top and average performers in the telephony industry.[17,18] Best in class companies, those in the top 20% of performance, cancelled the same percentage of projects overall, but were able to cancel unattractive projects much earlier in the process. This allowed them to focus their development resources on the truly important projects. They kept more of the bottom 50% of ideas from entering the development process; and as a result, completed projects in half the time and achieved double the percentage of revenue from new products, as compared to the other 80% of their competitors.

To squeeze everything you can through the bottleneck, there just isn't room for anything but the most promising projects. You need a uniform innovation process with an early assessment element to ensure that only the highest quality projects make it to your innovation bottleneck. Think of assessment as a hopper or funnel of potential projects at the front end of your innovation process. This hopper has a sieve at its exit that quickly diverts the losers and only releases potential winners to the constraint. Projects must be assessed for:

- Clear unmet or unarticulated market need
- Value of the solution to the customer
- Potential for a solution that can be delivered profitably
- Clear identification and evaluation of competitive alternatives

These steps help ensure that the constraint will not be in the marketplace and that the project has promise before committing expensive, constrained resources, such as development or testing. If the team has a compelling story consistent with company strategy, the next step is a project-planning proposal with the first step being to test feasibility.

You do have to be comfortable with some level of risk since feasibility resources are committed without a guarantee of success. Without some risk, there is a danger of cancelling projects too early, missing big ideas, and marginalizing product development to delivering only conservative line extensions. Look at feasibility as a small investment to help you understand whether the project risk is worth taking.

Stop bad multitasking

We regularly find that organizations struggle because they have key technical staff assigned to too many projects. Some companies' job postings even list multitasking as a skill requirement. These activity junkies have it upside down. The objective is not to work on more projects, but to actually complete projects that generate more throughput.

The bad multitasking we are concerned about here, though, is having the constraint switch back and forth between multiple projects without first completing an entire task. Studies show that productivity drops rapidly after engineers are assigned to more than two projects. By the time they are involved in 5 projects, less than 30% of their time is spent on value adding activity. You can dramatically improve results by limiting technical staff to a smaller number of projects and requiring them to complete tasks before switching projects.[13,29]

The general manger of a mid-sized industrial equipment company came to us with that very issue. We identified the bottleneck constraint in the engineering and development group, where they had far too many projects underway. They were consuming significant resources, but very few of them were making any progress. In fact, some of their best people were demoralized to the point where they were considering leaving.

We quickly whittled their active project list down to four projects, and the results were remarkable. With significant new visibility into each project, the teams had a new sense of purpose. They took a new technology project that had a three-year development horizon and, using a combination of already available technologies, began production of a new line of energy savings products with 95% of the benefits in less than 6 months. Eighteen months later, they reported that their average new product cycle time had dropped from 30 down to 10 months.

Guided Innovation Mapping

As teams begin planning a project, *Guided Innovation Mapping* is a particularly useful tool to ensure that they are working on the most important tasks. A variation on the TOC prerequisite-tree thinking tool, it identifies where innovations will be required and facilitates development of simple solutions. It also helps the team identify the critical, or hinge, assumptions they must consider for the project—these are the assumptions that define feasibility.

After project objectives are established, the project leader and cross-functional team participate in a facilitated session where each participant is asked, "What are all the possible obstacles that you could dream of that might cause this project to fail?" The simple act of asking people to provide input on why something might fail opens their minds to think in new ways and eliminates the passive aggressive behavior that can hurt team effectiveness. The wording of the question is important because we want them to be open and not to worry about being viewed as cynical or pessimistic. Of course, the team's next task is to develop a plan that gets around all of these obstacles and categorizes the resulting project steps as technical, commercial, manufacturing, regulatory or intellectual property.

Guided Innovation Mapping helps the team to understand where they must focus their innovation efforts. The hinge assumptions about key project constraints that come out of it are important for helping the team prioritize its efforts and define feasibility. Another benefit is that the team jointly develops a visual version of the project plan, including most of the inputs needed for Microsoft Project or other project planning tools.

One example is an industrial client that desired to begin moving some of its high-durability, high-efficiency technology out of the industrial market into a higher volume commercial market. The company had all of the technology required; they easily could have started down the development path without testing feasibility. However, through the process of *Guided Innovation Mapping*, we identified that commercial feasibility was a critical assumption that needed to be confirmed. With some quick competitive intelligence, they verified the unmet need for higher efficiency but found that customer payback requirements of 18-36 months resulted in a price point much lower than the team had anticipated. This forced the team to rethink their entire plan and refocus the early engineering effort on identifying actions they could take to simplify the option packages offered and value engineer out nearly 35% out of the cost of the product—a critical task defining feasibility for the entire project.

It may sound like common sense, but without disciplined planning and feasibility tools in place, companies can allow projects like this to get quite far along before identifying and dealing with potentially fatal flaws.

Critical Chain

Critical Chain applies Theory of Constraints to fix the issue of projects that persistently finish late—one of the most widespread frustrations in companies today.[20,23] Just as every process has a constraint, so does every project. The critical chain is the project's constraint, or the longest sequence of tasks comprising the project's cycle time or completion date, taking into account constrained resources. Because the critical chain is the constraint, it is also the leverage point for any improvement. We can only shorten the cycle time by shortening the critical chain.

1. Outline the sequence in which project tasks must be completed, either on paper or using project-planning software.
2. Resolve any resource contentions that have team members working on multiple tasks at the same time. Later, you may have to do the same thing for resources that are shared across multiple projects.
3. Work together with team members to estimate how long it will take to complete tasks with at least 90% cer-

tainty. Cut those estimates in half to obtain 50% certainty estimates.

4. Build the project timeline using the 50% estimates and then add in half of the task timeline at the end as a project buffer. No buffer is assigned to individual tasks.

5. If another sequence of tasks feeds into the critical chain, there should be a similarly constructed protective or feeding buffer at the end of that sequence to protect the critical chain.

6. The resource (people or equipment) assigned to each task must be ready to take the relay handoff as soon as the previous task is completed. Ideally, resource buffers can be added to make sure resources are freed up in advance and ready to start tasks without delay.

A simplified example is shown in figure 3.

Figure 3

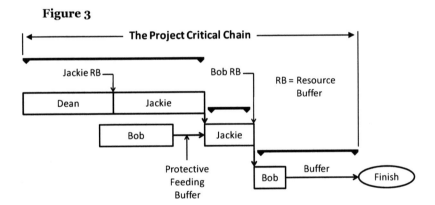

Prioritize the constraint

You can improve innovation throughput almost immediately by giving the best projects priority access to the bottleneck. Moving your bottleneck from producing 8 units per unit of time with a value of 5 each to producing 5 with a value of 10 each increases throughput by 25%. The key metric for prioritization is the expected return per unit of bottleneck time. This is loosely analogous to return on R&D investment.

It can be difficult for managers schooled in the paradigm of local optimization to accept, but only maximizing the return per unit of constrained resource will maximize the global return for the system.

Table 1 considers three projects for prioritization. If development time is the capacity constrained resource, project C would deliver $250 of projected return per hour of development, ranking it ahead of Projects B and A respectively. But if application testing is the shared resource constraint, project A only requires a moderate amount of bottleneck time and delivers $250 per hour of constraint. That ranks it highest for global system throughput ahead of projects B and C respectively.

Table 1

	Project A	Project B	Project C
Projected Return	$500,000	$350,000	$750,000
Development Time (hours)	4,000	1,500	3,000
Return per Development Hour	125	233	250
Ranking	3	2	1
Application Time (hours)	2000	1,500	4,000
Return per Application Hour	250	233	188
Ranking	1	2	3

Customer Value Lens:

Does your organization have a clear understanding of what customers value and what their unmet needs are? Do you know where their pain is, how to make them acutely aware of it, and how to position your new product as the solution? Or does your staff focus on the features of your products rather than the benefits that customers buy? Elevating these critical innovation and marketing abilities can pay big dividends.

Industrial B2B markets:

In a business-to-business setting, TOC provides a powerful lens for examining how you can add value for customers and how you communicate that value in your marketing and sales efforts. Since your customer's goal is also to make money now and in the future, the only way you can truly create superior value and induce them to buy is by helping them to make more money.

You can ensure that any innovation creates superior value by considering three elements known as ΔT, I, and OE, where the Greek letter delta (Δ) indicates the change or improvement that your innovation delivers: [16]

ΔT -   Increase in Throughput
ΔI -   Reduction in Inventory and Investment
ΔOE - Reduction in Operating Expenditures

Note that throughput (ΔT) is listed first for a reason. While maintaining a strong cost position is important, increases in throughput are bound only by your skill and creativity in finding new solutions and creating new market opportunities. How much more managerial effort do you think Apple puts on generating new throughput vs. cost control?

Market facing activities and customer visit work (different than sales calls) must be directed toward identifying unmet customer needs and then using the customer value lens of ΔT, I, and OE to understand the value created. When launching new products, you must communicate that value through both marketing and sales efforts, using the language of the various decision makers.[30] It's always about the customer's pain and how you help solve it. But at the operational level, it's about their unit cost targets and problems in meeting them; at the managerial level, it's about their productivity and budget problems; and at the top management level, it's about their problems with the bottom line - earnings or cash flow.

Consumer markets:

In consumer markets, we apply a different lens to customer value. The primary difference is in the definition of customer value. For consumers, convenience, emotion, and status-driven issues such as brand and fashion can play a much larger role. That's not to say that consumers don't buy on price, but they rarely conduct side-by-side economic analyses. The early popularity of Toyota's Prius hybrid attests to this with a pay-back period that was often longer than the expected life of the car. Nonetheless, the market for hybrids grew rapidly in large part due to the emotional connection that consumers feel with being green.

An alternative approach is to look at the jobs consumers are trying to do along with their key buying tradeoffs.[21] What are the tradeoffs along traditional and emerging competitive dimensions? Innovation then becomes a search for how these jobs can be done better along various lines - all as the basis for a new competitive dimension.

As an example, iRobot was able to identify vacuuming and floor washing as time-wasting jobs in the busy lives of today's families. The Roomba robotic vacuum and Scooba® floor cleaner automatically takes care of this job for consumers. How do they stack up along various competitive dimensions? They're good, but not nearly as effective or inexpensive as most alternatives. But iRobot isn't selling vacuums. It's selling the convenience of automated cleaning—a new dimension that has allowed them to bypass the market constraint of lower compet-itor pricing and establish a completely new market for auto-mated cleaning. Figure 4 below shows how Roomba® performs along these competitive dimensions.

Figure 4

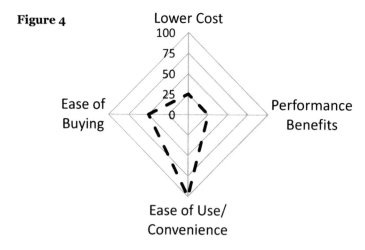

Innovation tools

Training and investment in innovation tools can help to develop strengths in the processes and skills necessary to elevate the innovation capacity of your organization.

Thinking processes for problem solving:

Inventions solve problems by resolving contradictions. Carbon fiber composites are a great example of an invention used to solve the contradiction between strength, weight, and speed. Nanotechnology promises to solve the contradiction between active material performance, high surface area requirements, and compact size. Invention and problem solving skills are an important component of innovation that can be exploited with training in various thinking processes. *Guided Innovation Mapping* is based on some of the thinking processes of TOC, which help to identify contradictions and find simple solutions that resolve them.[12,11] Altshuller's *Theory of Inventive Problem Solving* (TRIZ) is another tool that is growing in popularity.[8]

STEP 3. SUBORDINATE TO THE PACE OF THE CONSTRAINT

Now that you know what you need to do to fully exploit the bottleneck's capacity, how do you subordinate the organization priorities to the bottleneck and make sure that all of your non-constrained resources help in the effort to exploit it?

Bottleneck pulls from a prioritized buffer

How do you keep from distracting or interrupting the bottleneck with lower priority tasks and activities? Your project management system must ensure that the next highest priority task is always waiting and ready for the constraint. In a TOC manufacturing setting, the constraint pulls work from an inventory buffer. The same concept applies to innovation where you establish a buffer of prioritized tasks. In this way, the constraint does not waste any time deciding what to do next.

But how do we know what task the non-constrained resources should work on next. Because reducing cycle time is our goal in managing projects, the project with the highest percentage of buffer usage as compared to percentage of critical chain completion is most in danger of missing its target and should be the highest priority task.

The number of tasks waiting is set at a safety level to cover the normal fluctuations that you would expect in terms of the development group's work. The upstream groups continually monitor the buffer to make sure that they are timing release of their work to keep the buffer within the safety level.

In order to maintain this focus, your innovation process must have a clear communication mechanism. The process needs to have easy visibility into the status of projects and the prioritization buffer. Depending on the duration of projects, this can be as simple as daily five-minute team meetings, a regularly updated project intranet, or a more sophisticated software program.

Avoid early starts

Since non-constraints have extra capacity, why shouldn't they just get an early start on the next project? If we have the resources to do projects 1, and 2 simultaneously and projects 3 and 4 are already queued up in the prioritization buffer, how much harm is there in letting projects 5 and 6 get a head start, as well? After all, good research managers know that people need something to keep them busy during the inevitable breaks in a project. But remember, the non-constraints are working to keep the prioritization buffer at a target level, so we're talking about building an excess inventory of uncompleted tasks.

There is ample evidence from JIT and Lean to show that releasing work ahead of the required lead-time actually increases the lead-time required to complete all jobs.[16] What happens when the person that has gotten an early start on the number 5 project needs to jump back to help on the number 1 project? Do they immediately drop everything they were doing? Unlikely - when switching, they need time to finish up and document the lower priority task; this results in more frequent delays, decreased focus, and increased cycle time.

Help the bottleneck

Keeping non-bottleneck resources busy with additional projects doesn't help throughput or cycle time. However, an excellent use of their excess time is to unload the constraint by either helping with tasks or helping eliminate tasks. Let's take the case of a paint company with its bottleneck in performance testing. Anything the formulators can do to help or eliminate work for the testing group will reduce cycle time. This could be gathering all of the testing materials needed before the project is released to the constraint, helping in the testing lab, or doing additional journal research to see if similar work has been reported. It could even go as far as finding advanced experimental design techniques or modeling methods to reduce the number of samples required. By taking these steps, the formulators can take some of the load off the constrained testing group and help get more programs through the system in less time.

A client that was struggling with a product development project asked me to come in and help get things back on track. When we looked at all the work remaining, it appeared that the engineering group was the bottleneck. But a closer look revealed that other groups were better suited to some of the work the engineers were doing. Because one component of the project was a new control system, the engineers had excitedly started working directly with potential suppliers in negotiating the specifications and price for a solution. Clearly engineering input was required, but we were able to offload a significant amount of the control system sourcing to the supply chain group. It might seem like common sense in hindsight, but it could have gone unnoticed had we not been looking for ways to unload the constraint.

Using DOE to reduce work for the constraint:

If your constraint is in development or testing, your non-constrained resources should be using Statistical Design of Experiments (DOE) to reduce the number of experimental iterations sent through the bottleneck. These methods help researchers to create experimental plans with a minimum number of data points. JMP is one of the most widely used computer statistical packages for DOE.[31]

Cross-functional engagement

After all of the effort expended to exploit the constraint, the last thing you want to do is allow delays or mistakes downstream of the constraint to reduce throughput. Returning to our paint example, what sense would it make to develop and test a new paint if they weren't confident that they could scale it to production or that they had selected the right channel to bring it to market? Your innovation process must subordinate resources across the organization to be involved in the innovation process. Downstream groups should be involved early enough and frequently enough to ensure that they don't see any insurmountable hurdles and are ready to take the project forward as soon as the constraint completes its work.

STEP 4. ELEVATE THE CAPACITY OF THE CONSTRAINT

After subordinating everything else to your plan for exploiting the constraint, what can you do to elevate or increase the capacity of your bottleneck innovation resources? So far, the first three steps have been focused on eliminating policies and changing the way you operate to make the constraint operate as efficiently as possible. The improvements, while not necessarily easy, should have been rapid and delivered a high return on investment. The elevate step starts to ratchet up the investment (training, partnerships, staff, equipment, etc.) so you should not proceed until you have exhausted steps 2 and 3.

Let's look at some of the levers that you can pull to elevate the capacity and capability of innovation:

Fundamentals Research, modeling and simulation:

DOE is a useful empirical method to better exploit the bottleneck, but it will only take you so far. Conducting fundamentals research, either internally or externally, can give you an important understanding that actually adds to your innovation capacity. Computer modeling and simulation goes a step further, by taking that fundamental knowledge and allowing researchers or designers to conduct virtual experiments and play 'what if' simulations to predict performance.

Employee development & selection

Having top talent is important throughout the organization, but nowhere more than in the group entrusted to making money in the future. The product managers, engineers, scientists, and technologists in these roles can have a profound impact on your success.[27,26] It is crucial to have regular development discussions with your people—including clear feedback on their skills and competencies and whether they have the ability to become top talent in their current roles. If they don't, then you need to consider how you can move top talent into those roles without disrupting the organization.

Open innovation

One of the single largest constraints in any organization is simply that there are only so many managers, so many funds to invest, and so many people to carry out the work. One solution to elevating your capabilities is open innovation. By reaching outside the walls of your own company to bring in technology or market opportunities through alliances, you can access the innovation capabilities of a wide array of external players – other companies, research firms, start-ups, universities, entrepreneurs, etc. On the other hand, if you have the technology but don't have the channel to market, you can extend your reach through licensing or marketing alliances.

Before you get too excited and run out to put together that alliance, you should know that nearly 70% of alliances fail.[28] Why such a high percentage? Because people often get caught up in the excitement of the deal and don't pay attention to the fundamentals of strategic alignment between the partners. Alliances add a level of complexity that you must be ready for and should only be given serious consideration after you have attained some level of discipline and capability with your own projects. When you are ready, Sagal and Slowinski's Alliance Framework® and WFGM Model® provide a solid framework for the four competencies you need to successfully implement OI:[32]

- Want - Defining the outside capabilities that you need
- Find - Finding potential partners that have those capabilities

- Get - Using proven process for partner selection and alliance negotiation
- Manage – Implementing and managing the alliance

STEP 5. START AGAIN AND AVOID THE INERTIA CONSTRAINT

As you drain the swamp, more alligators will surface. As soon as one innovation constraint is broken, another will show up. It's time to start the cycle again and refocus another round of improvement efforts on the next bottleneck. If you stop after a single round, then inertia becomes your constraint and you only end up stagnating at a higher level. That's the nature of continuous improvement. Identify, exploit, subordinate, elevate, and do it all over again.

Of course, there are always exceptions, and sometimes it is best to manage your constraint, rather than to break it. In this case, as you continue to drive improvement and elevate your constraint, the fifth focusing step becomes regular monitoring and improvement of the non-constraints to maintain the protective capacity needed to ensure steady flow to the bottleneck. An example might be in fundamental research and new technology development where you are creating new capabilities that your product development group will need in the future.

Before proceeding with the next round, you also should examine whether the policies that governed the old bottleneck still apply. For example, a policy that certain members of the development team were always kept 100% busy doing development work in the lab might have made perfect sense when the bottleneck was in development. However, if that constraint is broken and getting enough new ideas into the pipeline is the new bottleneck, that old policy might now be a constraint. Perhaps getting those development people out into the field to understand unmet customer needs would better serve your goal of making more money in the future. Similarly, you should consider whether the previous metrics are still appropriate.

Putting it all to work

A poet once said:

Knowing is not enough; we must apply.
Willing is not enough; we must do. [33]

So, as leaders, how do you begin to put this approach into practice? It would be easy to say that it's just a matter of discipline in identifying your constraint and then taking the new product and service development activity in your business through the TOC focusing steps. But we all know change is more difficult than that. The unknown elements of change can be a threat to security, causing fear and resistance. Knowing that, there are multiple layers of resistance that you must be prepared to take your organization through: [34]

Layer 1: Has the right problem been identified?

Layer 2: Is this solution leading us in the right direction?

Layer 3: Will the solution really solve the problems?

Layer 4: What could go wrong with the solution? Are there any negative side effects? How can we avoid them?

Layer 5: Is this solution implementable? What is our plan?

Layer 6: Are we all really up to this? Is management committed to ongoing innovation improvement or is this just another program that will come and go?

These are normal concerns your people might have. Take measured steps, one focusing step at a time, and most importantly, involve them. Invest in a coach and train your people in the concepts of TOC applied to innovation. Ask them what the obstacles are to using it effectively in your organization, and then involve them in finding solutions around those obstacles. Some of the tools we've already discussed, such as innovation mapping and the TOC thinking tools, can be very helpful for this purpose. It is critical that people feel secure in expressing their opinions and issues without having it come back to haunt them. A facilitator can help with the process, but as the leader, you must also be involved to establish trust and help bring people through the layers of resistance.

You'll find that your people develop common sense solutions. Maybe even more importantly, you'll find that their ownership of the solution creates the commitment needed to break the constraints holding back your growth and achieve your goals of making more money in the future.

Afterword – Your Next Steps

"Simplicity is the final achievement. After one has played a vast quantity of notes and more notes, it is simplicity that emerges as the crowning reward of art."

- Frederic Chopin (1810 – 1849)

Well, there you have it—the five steps that you can take to begin seeing more bottom-line impact from the innovation resources you already have. We've tried to include the best and latest thinking on the subjects of innovation and marketing,—all within the framework and thinking processes of TOC as we've presented via Maggie's story. While it's impossible to cover all the possibilities in a single volume that's still easy to slip into a briefcase or purse, what we haven't been able to include here is available to you in the growing collection of resources at:

www.SimplifyingInnovation.com/extras

Now, some may question whether what we've presented here is nothing more than common sense. It might be, but I think you'll agree that it's simply not the way that most companies operate today. Perhaps, the more important question is, now that you've been exposed to these concepts, will you pass them over as obvious? Or will you view them as common sense that isn't commonly practiced? Most importantly, will you begin profiting from them by systematically using the five steps to achieve more profitable growth with the product development resources you already have?

Since developing this framework, we've helped companies use these methods to cut time to market by more than half, nearly double new product profits, and get more results than they ever thought possible from their new product development investments.

It is our hope that you will recognize the benefits for your company and begin to put this high-leverage approach to work for yourself and reach more of your goals in doing so.

Simplifying Innovation gives you the roadmap needed to begin using the Guided Innovation System. But should you ever need assistance navigating during your journey, our coaching and consultation can guide you in the same way it has others.

And as the poet once said...

Knowing is not enough; we must apply.
Willing is not enough; we must do.

Michael A. Dalton
Guided Innovation Group

Reference Notes:

1   Boston Consulting Group. 2006. *Innovation 2006. Boston, MA.*

2   In 2003 the $4.2 trillion dollar US Manufacturing economy spent $150 billion (3.6%) on R&D.

3   Jaruzelski, Barry, Kevin Dehoff and Rakesh Bordia. Booz Allen Hamilton *Global Innovation 1000: Money isn't everything*, Strategy+Business, Issue 41, Winter 2005

4   While often used as a generic term for gated product development processes, Stage-Gate is a trademark of Stage-Gate International.

5   Schragenheim, Eli and H. William Dettmer. 2001. *Manufacturing at Warp Speed: Optimizing supply chain financial performance*, Boca Raton, FL: St. Lucie Press.

6   Pirasteh, Reza M. Kimberly S. Farah. *Continuous improvement trio: the top elements of TOC, lean, and six sigma make beautiful music together*, APICS magazine, May, 2006

7   DeBono, Edward. 1973. *Lateral Thinking: Creativity Step by Step*, New York, NY: Harper Colophon

8   Altshuller, Genrich. 1999. *Innovation Algorithm: TRIZ, Systematic innovation and technical creativity*, Worcester, MA: Technical Innovation Center, Inc.

9   Goldratt, Eliyahu M. and Jeffrey Cox. 1984. *The Goal*, Great Barrington, MA: North River Press

10 Ohno, Taiichi and Norman Bodek, 1988. *Toyota Production System: Beyond Large-Scale Production*, New York, NY: Productivity Press.

11 Dettmer, H. William, 2007. *The Logical Thinking Processes: A systems approach to complex problem solving (second edition)*. Milwaukee, WI: American Society for Quality

12 Goldratt, Eliyahu M. 1994. It's not Luck. Great Barrington, MA: North River Press

13 Wheelright, Steven C. and Kim B. Clark, 1992. *Revolutionizing product development: Quantum leaps in speed, efficiency, and quality*. New York, NY: Free Press

14 Hlavacek, James D. 2002, *Profitable Top-Line Growth for Industrial Firms*, Knoxville, TN: American Book Co.

15 Christensen, Clayton M. and Michael E. Raynor. 2003. *The Innovators Solution*, Boston, MA: Harvard Business Press

16 Goldratt, Eliyahu M. 1990. *What is this thing called theory of constraints and how should it be implemented*. Great Barrington, MA: North River Press

17 Eduardo, Miranda, 2003. *Running the Successful Hi-tech Project Office*, Norwood, MA: Artech House

18 Ogawa, Dennis and Laura Ketner. January 27, 1997. *Benchmarking product development*. Telephony Online, Chicago, IL: Penton Media

19   When cash for investment is a constraint, which it usually is, TOC also offers an approach called dollar days that integrates the investment and the time duration that it is tied up vs. how long it takes the sales throughput dollar days to offset the investment dollar days. This measure is also known as the time to reach flush.

20   Goldratt, Eliyahu M. 1997. *Critical Chain*, Great Barrington, MA: North River Press

21   Mankin, Eric, 2004. *Can you spot the sure winner?* Strategy and Innovation: Volume 2, Number 4. Boston, MA: Harvard Business Press

22   Koch, Richard. 1998. *The 80/20 Principle*, New York, NY: Doubleday

23   Leach, Lawrence. 2005. *Lean Project Management*, Boise, ID: Advanced Projects, Inc.

24   Anthony, Scott D. and Clayton M. Christensen. *Building Your Growth Engine*. Strategy & Innovation: Jan-Feb 2005. Boston, MA: Harvard Business Press

25   Some have claimed that this is more appropriately Exploit; I would argue that without first subordinating the non-capacity constrained resources we would lack the time that is required to help the bottleneck.

26   Buckingham, Marcus and Donald O. Clifton. 2001. *Now Discover Your Strengths*, New York, NY: Free Press

27   Smart, Bradford D. 2005. *Topgrading: How leading companies win by hiring, coaching, and keeping the best people, revised and updated edition*, New York, NY: Penguin Group

28   Slowinski, Gene and Matthew W. Sagal. 2003. *The Strongest Link*, New York, NY: AMACOM

29   Rubinstein, Joshua S. and David E. Meyer. *Executive control of cognitive processes in task switching*, Journal of Experimental Psychology - Human Perception and Performance, Vol. 27, No.4

30   Goldratt, Eliyahu M. Eli Schragenheim and Carol A. Ptak. 2000. *Necessary But Not Sufficient*, Great Barrington, MA: North River Press

31   *JMP Statistical Discovery*. Cary, NC: SAS corporation

32   Slowinski, Gene. 2005. *Reinventing Corporate Growth*, Gladstone, NJ: Alliance Management Group

33   Johann Wolfgang von Goethe. German Poet, 1749-1832

34   The Goldratt Institute, 2001. *The theory of constraints and its thinking processes*

Acknowledgements

First, I thank Carol, my wife, for standing by me with a joyful heart and sense of perspective, and most of all for loving me. You've been amazingly understanding, not just during this project, but throughout all our years together.

I thank my parents, Norman and Freda, for teaching me how to make things work and get things done and for raising me to be a believer. I only wish you could have still been here, Dad. Special thanks to Erwin, Jean, Beth, Peter, Andy, and Brenda for all your love, even when I was distracted with other things.

I also thank the folks who have been my friends and mentors in the development of this book:

To Dr. Paul Gloor of BASF Corporation, a friend and colleague who introduced me to TOC and initially encouraged me to explore how it could become a framework for driving new product improvement.

To David Brotton and Dr. Rich Chylla, friends and colleagues who are always eager to explore new approaches.

To Dr. James Hlavacek, CEO of the Corporate Development Institute, who taught me how to uncover what customers need.

To Dr. Gene Slowinski and Dr. Matt Sagal of the Alliance Management Group, who taught me how to help companies play together profitably.

To Doreen Miller at iMagi-Media & Design, who artistically helped me put my vision online and into print.

To my editor, Patti McKenna, who helped give my writing added clarity, warmth and depth.

To my friends at EA and the PEN II group, who are always a reliable sounding board and source of encouragement.

Of course, many colleagues were amazingly generous with their time and energy in the review process, helping to make this a better book. My sincere appreciation to H. William Dettmer, Carol Ptak, Richard Klapholz, Dr. Lisa Lang, Guillermo Cabiro, Bud Equi, Dr. Marco Villalobos, Bill Blasius, Beth Robinson, Drea Knufken, Dr. Victor Newman, Lynn Dessert, Pradeep Henry, and Jeffery Baumgartner.

I'd especially like to thank my clients over the years—for entrusting me to help you with your struggles and for all you have allowed me to learn from your organizations.

And of course, I thank the Lord, through whom all things are possible.

Index of Key Concepts

About the Author

Mike Dalton is the founder of Guided Innovation Group, whose simple mission is helping companies turn their new product innovation into bottom-line impact. Guided Innovation's unique TOC-based approach to rapid innovation improvement is helping companies slash time to market in half and nearly double new product profits.

Mike's new product innovation experience started with over 24 years of executive management leadership at the SC Johnson family of companies, the multi-billion dollar, privately held consumer and industrial products multinational. He has grown new and existing businesses as a general manager and in marketing, business development, and manufacturing leadership roles. He holds an MBA in marketing and finance from the University of Chicago and a degree in chemical engineering and energy technology from the Illinois Institute of Technology. He speaks on innovation and growth for trade and executive groups and has spoken on Open Innovation for the National Academy of Sciences. He lives in Mt. Pleasant, Wisconsin with his wife, Carol, and their rescue pitbull terrier.

Whether you are struggling to get more sales impact from your new product and innovation investment or are growing strongly but still interested in taking your innovation performance to the next level, we welcome your call.

Find out more at www.GuidedInnovation.com
Or contact Mike at (262)672-2700

Small Changes Can Deliver Big New Product Results

Sign up for your free Breakthrough Strategy Session. You'll identify the issues that are holding your company back and learn powerful strategies for increasing your growth. To take that next step, simply visit our website where you'll find the form below.

Yes, I'm ready to take you up on your free offer!

Please have a Guided Innovation representative contact me by telephone to schedule my free Breakthrough Strategy Session.

My Key Frustrations Are:

☐ Finding new product opportunities   ☐ Project costs coming in over budget
☐ Developing robust solutions   ☐ Funding projects that get canceled
☐ New product sales below budget   ☐ Lack of focus on important projects
☐ New product projects finishing late   ☐ People don't seem to be engaged
☐ Staff saying they're stretched too far ☐ No framework for improvement

I understand that by participating, I am not obliged to buy anything and that my information will be held in strict confidence.

Here's all you need to do:

Simply fill out this form online at:
http://www.GuidedInnovation.Com/Breakthrough

Phone us at 1-262-672-2700

Or email us at Breakthrough@GuidedInnovation.com

CPSIA information can be obtained at www.ICGtesting.com
Printed in the USA
LVOW08s2238060215

426021LV00032B/2598/P

9 780615 329390